Contemporary Trends and Issues in Science Education

Volume 54

The book series Contemporary Trends and Issues in Science Education provides a forum for innovative trends and issues impacting science education. Scholarship that focuses on advancing new visions, understanding, and is at the forefront of the field is found in this series. Authoritative works based on empirical research and/or conceptual theory from disciplines including historical, philosophical, psychological and sociological traditions are represented here. Our goal is to advance the field of science education by testing and pushing the prevailing sociocultural norms about teaching, learning, research and policy. Book proposals for this series may be submitted to the Publishing Editor: Claudia Acuna E-mail: Claudia.Acuna@springer.com

Christine D. Tippett · Todd M. Milford
Editors

Seeing Science Through the Eyes of Canadian Teachers and Learners

Research in Secondary and Tertiary Science Education

Springer

Editors
Christine D. Tippett
University of Ottawa
Ottawa, ON, Canada

Todd M. Milford
University of Victoria
Victoria, BC, Canada

ISSN 1878-0482 ISSN 1878-0784 (electronic)
Contemporary Trends and Issues in Science Education
ISBN 978-3-031-88784-0 ISBN 978-3-031-88785-7 (eBook)
https://doi.org/10.1007/978-3-031-88785-7

This Springer imprint is published by the registered company Springer Nature Switzerland AG
The registered company address is: Gewerbestrasse 11, 6330 Cham, Switzerland

Acknowledgements

This book would not exist without the help and support of many individuals. We would like to thank contributing authors, reviewers, the editorial team at Springer, as well as science teachers and science education professors across Canada.

We also want to acknowledge by name Dr. Larry D. Yore, our wise and patient mentor who continues to support us without question (well maybe a few questions), and Shari Yore, whose careful reading of the manuscript, combined with her grammatical expertise, was invaluable.

Christine D. Tippett
Todd M. Milford

List of Reviewers

David Blades, University of Victoria
G. Michael Bowen, Mount Saint Vincent University
Xavier Fazio, Brock University
Victor Glickman, University of British Columbia
Michael Hammond-Todd, Utah Tech University
Darron Kelly, Memorial University of Newfoundland
Alena Kottova, University of Victoria
Wendy L. Kraglund-Gauthier, Yorkville University
Mark McDermott, University of Iowa
Katarin A. MacLeod, St. Francis Xavier University
Anik Meunier, Université du Québec à Montréal
Wayne Melville, Lakehead University
Michael-Anne Noble, Royal Roads University
Natascia Petringa, University of Ottawa
Sandra Scott, University of British Columbia
Astrid Steele, Nipissing University (emeritus)
Ellen Watson, Brandon University
Dawn Wiseman, Bishops University
Elizabeth Vergis, St. Mary's University
Grant Williams, St. Thomas University
Larry Yore, University of Victoria (emeritus)

Contents

Editors and Contributors

About the Editors

Christine D. Tippett, B.A.Sc. (University of British Columbia), B.Ed. (University of Victoria), MA (University of Victoria), Ph.D. (University of Victoria), is a full professor of science education in the Faculty of Education at the University of Ottawa. She was an engineer before she obtained her teaching degree, which influences her ways of thinking about science education. Her research interests include visual representations, science education for all students, and professional development for science educators (pre-service, in-service, and informal). Current projects focus on preservice science teachers' images of engineers, early childhood STEM education, and assessment of representational competence.

Todd M. Milford, B.Sc. (University of Victoria), B.Ed. (University of Victoria), Dip Spec Ed (University of British Columbia), M.Ed. (University of Victoria), Ph.D. (University of Victoria), is an associate professor and Acting Dean in the department of Curriculum and Instruction at the University of Victoria. Prior to this he was a lecturer in the Art, Law, and Education Group at Griffith University in Brisbane Australia. He has science and special education classroom teaching experience as well as in the on-line environment. He has been teaching at the postsecondary level since 2005 primarily in the areas of science education, mathematics education, and classroom assessment. His research has been and continues to be varied; however, the constant theme is using data and data analysis to help teachers and students in the classroom.

Contributors

Anthony W. Bartley Lakehead University, Thunder Bay, ON, Canada

Charlène Bélanger Planetarium de Montréal, Montreal, QC, Canada

G. Michael Bowen Faculty of Education, Mount Saint Vincent University, Halifax, NS, Canada

Pei-Ling Hsu University of Texas - El Paso, El Paso, TX, USA

Darron Kelly Memorial University, St. John's, NL, Canada

Wendy L. Kraglund-Gauthier Yorkville University, Fredericton, NB, Canada

Katarin A. MacLeod St. Francis Xavier University, Antigonish, NS, Canada

Wayne Melville Lakehead University, Thunder Bay, ON, Canada

Anik Meunier University of Quebec in Montreal (UQAM), Montreal, QC, Canada

Todd M. Milford University of Victoria, Victoria, BC, Canada

Michael-Anne Noble Royal Roads University, Colwood, BC, Canada

Sharon Pelech University of Lethbridge, Lethbridge, AB, Canada

Jrène Rahm Université de Montréal, Montreal, QC, Canada

Latika Raisinghani University of Winnipeg, Winnipeg, MB, Canada

Giuliano Reis University of Ottawa, Ottawa, ON, Canada

Joy Ricketts Roncalli Central High, Avondale, NL, Canada

Wolff-Michael Roth University of Victoria, Victoria, BC, Canada

Christine D. Tippett University of Ottawa, Ottawa, ON, Canada

Elizabeth Vergis University of Alberta, Edmonton, AB, Canada

Ellen Watson Brandon University, Brandon, MB, Canada

Patrick R. Wells Memorial University of Newfoundland, St. John's, NL, Canada

Dawn Wiseman Bishop's University, Sherbrooke, QC, Canada

List of Figures

List of Tables

Chapter 1
Recognizing—and Reframing—Secondary and Tertiary Science Education Trends in Canada

Todd M. Milford and Christine D. Tippett

1.1 Introduction

Canadian students consistently perform exceptionally well in international science assessments like the Organisation for Economic Co-operation and Development (OECD) Programme for International Student Assessment (PISA). For example, in the 2022 iteration, Canada ranked seventh in science out of 81 participating OECD countries and economies (OECD, 2023). In the 2018 iteration, Canada ranked sixth in science out of 79 participating OECD countries and economies (OECD, 2018). It is clear that Canadian students have been identified as top science achievers globally. However, despite these impressive outcomes, there has been limited systematic examination of the factors contributing to Canadian students' success in science education. To address this gap, we have been working to bring together research that explores the unique features to explain the effectiveness of science education across Canada's elementary, secondary, and tertiary levels, providing valuable insights into the systems and practices that underpin these strong international results. This book is the third in a series of edited volumes on Canadian science education, curriculum, instruction, and research. Across all volumes, we adopted a micro-meso-macro analytical lens (Dopfer et al., 2004; Serpa & Ferreira, 2019) that would highlight aspects of science curriculum, teaching, learning, and research in Canada through examination of micro-level, contextually constrained studies that can facilitate macro-level systemic comparisons when meso-level factors are considered.

T. M. Milford (✉)
University of Victoria, Victoria, BC, Canada
e-mail: tmilford@uvic.ca

C. D. Tippett
University of Ottawa, Ottawa, ON, Canada
e-mail: ctippett@uottawa.ca

C. D. Tippett and T. M. Milford (eds.), *Seeing Science Through the Eyes of Canadian Teachers and Learners*, Contemporary Trends and Issues in Science Education 54,
https://doi.org/10.1007/978-3-031-88785-7_1

The first volume, *Science Education in Canada* (Tippett & Milford, 2019), documented the commonalities and diversities in science education across the 13 provincial and territorial jurisdictions as well as detailed specifics of school organization, curriculum, teacher education and certification, professional development, and jurisdictional context. Canada's geographical, historical, and cultural considerations emerge from the complexities of its large landmass, unequal distribution population, French and English language requirements, Indigenous Peoples rights and traditions, and immigration patterns. These factors contribute to diversity in our science education systems, including schools, curriculum requirements, and teacher education. The 2019 volume detailed the provincial and territorial specifics of school organization, curriculum, teacher education, certification, professional development, and jurisdictional challenges. It also presented the meso-level consistencies, commonalities, and distinctions that are needed to better understand micro-level science education research across the country and to make macro-level comparisons internationally.

The second volume, *Exploring Elementary Science Teaching and Learning in Canada* (Tippett & Milford, 2023a), presented a micro-level perspective, exploring research in elementary schools. An analysis of the chapters included in this volume revealed significant regional variations and systemic challenges. While a national structure exists in the *Common Framework of Science Learning Outcomes, K to 12* (Council of Ministers of Education, Canada [CMEC], 1997), provinces and territories demonstrate distinct approaches to science curriculum, with notable differences in Indigenous knowledge integration, instructional practices, and resource allocation. Key issues uncovered at a micro-level include low prioritization of science education, inadequate teacher preparation—particularly at the elementary level—and linguistic barriers that complicate science learning for diverse student populations. The 2023 volume highlighted the complex interplay between national educational guidelines and local contexts, underscoring the need for more comprehensive and culturally responsive science education strategies that address the unique challenges across Canada's diverse educational landscape.

Taken together, the two volumes provide a foundation to understand contemporary explorations in secondary science education research. This third volume extends our exploration, with a focus on Canadian secondary and tertiary science education research. The collected chapters highlight contemporary and emerging issues such as environmental education, Indigenous perspectives, and innovative assessment practices, providing snapshots of the diverse mosaic of Canadian science education. With chapters exploring culturally responsive teaching, environmental consciousness, and progressive pedagogical approaches, the volume showcases how Canadian researchers and educators are addressing global challenges through localized and innovative practices. While firmly rooted in the Canadian context, the current and two earlier volumes offer insights into teaching and learning as well as research methodologies that will resonate with international audiences, making it a valuable resource for understanding current trends and future directions in science education.

1.2 Canada's Science Education System

Canada's education system is characterized by its decentralized governance; each of the 13 provinces and territories maintains distinct yet similar elementary, secondary, and tertiary educational systems that are influenced by local geographical and cultural contexts (Tippett & Milford, 2023b). Historically rooted in missionary education and reflecting linguistic diversity, curriculum can be offered in English and/or French and occasionally Indigenous languages. Overseen by provincial and territorial ministries, the system encompasses Kindergarten through Grade 12 education (typically for students ages 5 to 17 years) with varied grade configurations and school structures. While each jurisdiction is unique, the CMEC routinely brings together all education ministers to discuss assessment of student performance, develop learning outcomes and indicators, and promote priorities such as Indigenous education (CMEC, n.d.-a).

Across all Canadian jurisdictions, science is a mandatory subject from Kindergarten to Grade 10. Science curriculum across the country has been influenced by the *Common Framework* (CMEC, 1997) that was collaboratively developed by government officials, science consultants, and teachers. This document established foundational principles for scientific literacy and suggested grade-level learning outcomes, which individual jurisdictions have interpreted and implemented to reflect local needs. Recent examinations of the *Common Framework* have revealed the need for greater consideration of science, technology, society, and environment and contemporary topics and issues that were not foreseen in the 1990s.

Typically, elementary science curricula for Kindergarten to Grade 7 include general science topics covering the natural, physical, earth, space, and environmental sciences. Elementary teachers tend to be generalists with limited science backgrounds (Milford & Tippett, 2019). Conversely, secondary teachers are usually specialists who have completed multiple, related university courses; for science teachers these courses would be related to one or more disciplines (Milford & Tippett, 2016; Tippett & Milford, 2017). In Grades 8–10, science courses maintain a general science structure, progressing to discipline-specific courses such as biology, chemistry, and physics in Grades 11 and 12. Upper secondary students (i.e., Grades 11 & 12) can select science courses based on their personal interests, academic performance, and future educational or career objectives. Course selection is crucial because it can determine potential academic pathways, including university, college, or direct entry into the workforce. While graduation from secondary school typically requires the completion of one Grade 11 science course, students who intend to pursue tertiary education in scientific or technical fields often complete several upper-level science courses. Course selection at this level can be influenced by university prerequisites, such as Physics 12 for engineering courses. Locally developed secondary science courses, such as environmental education and experiential land-based options, have slowly gained recognition as accredited prerequisite science courses by some postsecondary science and engineering faculties. This growing recognition has opened the door for the development of additional innovative courses, including those focused on Indigenous ways of knowing and traditional ecological knowledge and wisdom, which further enrich science education opportunities in Canadian secondary schools.

Canada's tertiary education system includes 110 degree-granting universities (Statistics Canada, 2024a), approximately 150 recognized public colleges, which may also include "specialized institutes, community colleges, institutes of technology, colleges of applied arts and technology, or cégeps [collège d'enseignement general et professionnel/general and vocational colleges]" (CMEC, n.d.-b, Postsecondary Education section), and several thousand private colleges. In 2023, 26% of Canada's tertiary-level students were enrolled in science, mathematics, and engineering related programs (Statistics Canada, 2024b). Science education at the tertiary level is typically highly siloed, with the disciplines of biology, chemistry, and physics being further separated into specialty courses such as cell biology, organic chemistry, and quantum physics. While science professors are content area experts, they may have little or no formal pedagogical education. Many universities recognize this problem and are beginning to offer pedagogy-focused workshops for faculty members and graduate students who want to improve their teaching expertise.

1.2.1 Science Teacher Education

To situate the research presented in this book, it is helpful to know a little more about science teacher education in Canada. Teachers in publicly funded schools must hold a bachelor's degree in education—obtained by completing an approved teacher education program—as well as a teaching certificate. These programs, both elementary and secondary, share a common overarching structure. Admission requirements typically include minimum grade standards, completion of undergraduate coursework, relevant experience, and language proficiency. Universities in Canada are recognized as having higher admission requirements than universities in many other countries (Crocker & Dibbon, 2008). Once admitted to a teacher education program, students take educational foundations, learning exceptionalities, psychology, and methods courses, among others. They also complete one or more practicum placements. Programs reflect jurisdictional policies and preferences as well as individual institutional contexts, resulting in variations in course requirements and implementation.

Secondary science teacher education programs vary by institution. However, there are three main options: direct entry programs of 4–5 years after high school; post-degree programs of 8–20 months that can be taken upon completion of a university degree; or concurrent programs of 4–6 years that result in a science degree and an education degree (Olson et al., 2015; Tippett & Milford, 2017). Most universities have stricter requirements for secondary science education teachers than for elementary generalist teachers who are expected to teach science (Tippett et al., 2019a, 2019b). Preservice teachers are required to complete methods courses ranging from general science teaching approaches to discipline-specific instructional methods like secondary chemistry (Tippett & Milford, 2017). Secondary science education programs frequently allow for specialization in biology, chemistry, or physics, and occasionally science, technology, engineering, and mathematics (STEM). Classroom experience is a key component of secondary science teacher education programs, with 14- to 24-week practicum placements that provide teaching experiences in different school settings.

1.2.2 Secondary Science Teaching

Although preservice teachers receive extensive theoretical preparation, they often find a disconnect between academic learning and practical teaching experiences (Milford & Tippett, 2016). The challenges that Canadian secondary science teachers encounter in their professional practice reflect the rapidly evolving demands of education in the twenty-first century (MacDonald, 2010). The integration of technology in classrooms requires teachers to adapt their instruction using new and often unfamiliar digital tools (Walan, 2020). Teachers are increasingly being asked to incorporate culturally relevant science curriculum in efforts to meet the needs of diverse students and to enhance Indigenous students' educational opportunities (Kim, 2017). Many teachers face the challenge of differentiating instruction for progressively more diverse classrooms, including English language learners and students with special educational needs, all while maintaining rigorous academic standards (Maeng & Bell, 2015). These challenges demand ongoing professional development, resource investment, and systemic support to enable science teachers to effectively meet the needs of their students. Fortunately, Canadian secondary science teachers seem to be well supplied with materials and resources (Milford & Tippett, 2016). Additionally, there is an active community of science education researchers who engage in explorations of effective teaching and learning at all grade levels and across a range of contexts.

1.3 About This Book

This book emerged from our recognition of the need to document and update our understanding of the diverse landscape of Canadian science education research. We envisioned a collection that would elaborate on insights from *Science Education in Canada* (Tippett & Milford, 2019) and provide a snapshot of current and emerging research trends in science education.

We issued a call for submissions through the Science Education Research Group (SERG), the major organization for Canadian science education researchers, which generated high interest from our colleagues across the country and yielded more than 20 expressions of interest. Based on the response, we decided to craft two books: one exploring science teaching and learning in elementary settings—*Exploring Elementary Science Teaching and Learning in Canada* (Tippett & Milford, 2023a)—and one in secondary and tertiary settings.

This book contains an introduction and an epilogue that bookend a collection of 11 chapters presenting micro-level descriptions of science education research conducted in secondary and tertiary settings. These chapters represent a range of geographic settings, grade levels, student or teacher foci, topics, and methods (see Table 1.1). The research was conducted in settings including British Columbia, Nunavut, and Newfoundland and Labrador—truly reaching from sea to sea to sea. Eight chapters are situated in Grades 7–12—the full range of secondary education, and three chapters explore tertiary science education at universities or colleges. Eight chapters focus on teachers, while the remaining three chapters focus on students. In contrast to the chapters in the elementary research book (Tippett & Milford, 2023a), none of the chapters in this book focus on teachers and students together, even when the research was conducted in a classroom context. Teacher education or professional development remained an important theme, with four chapters exploring aspects such as learning to teach during COVID-19, action research during the design of novel ocean education activities, developing meaningful formative feedback, and changes in teaching practices during a 2-year museum collaboration. Overwhelmingly, our contributing authors reported on qualitative research, with nine chapters describing case studies, discourse analysis, narrative reflection, and phenomenology based upon artifacts, interviews, narratives, observations, and videorecording and two chapters describing mixed methods approaches.

1.4 Trends in Science Education Research in Canada

In this section, we demonstrate how the secondary science education research reported in this book aligns with, or differs from, national and international research topics as shown in our master list (see Table 1.2). This master list began with nine topics that have been used in content analyses of three international science education journals between 1998 and 2017: teacher education; teaching; learning (conceptions); learning (contexts); goals and policy, curriculum, evaluation, and assessment; culture, social, and gender issues; history, philosophy, epistemology, and nature of science; educational technology; informal learning (Lin et al., 2018; Tsai & Wen, 2005).

Next, we added the top ten topics presented between 2013 and 2017 at SERG's annual conference: teaching, STEM; inquiry and other skills; preservice teachers; environmental issues; science content knowledge; student focused; identity and self-efficacy; professional development; social justice and diversity (Tippett et al., 2017). To complete our master list, we added the themes that emerged from our first book on Canadian science education: a low priority of science; influences on science curriculum; Indigenous perspectives; inadequate science teacher education; language

Table 1.1 Overview of chapters

Chapter	Geographic setting[a]	Grade level	Focus		Topic	Method[b] (Data source)
			Student	Teacher		
2	SK	Grade 12		☆	Curriculum change, teacher concern, Concerns Based Adoption Model	QUAL (interviews)
3	NS	Secondary science methods		☆	Pedagogical awareness, emergency preparedness, creativity in teacher education	QUAL (narrative reflections)
4	BC	Grades 9–12		☆	English Language Learners, (trans-multi) culturally responsive education	QUAL (interviews, classroom observations)
5	NL	Grade 10		☆	Ocean education, pedagogical content knowledge	QUAL (lesson plans, classroom observations, reflections)
6	NB	Secondary science methods		☆	Assessment, formative feedback, laboratory reports	QUAL (video recordings, preservice teacher artifacts)
7	BC	Grade 11		☆	News media, nature of science, discourse analysis	QUAL (interview, video recording)
8	AB	Undergraduate chemistry	☆		Science knowledge, concepts of evidence, adapted primary literature	MM (pre- and posttests, interviews)
9	BC	Undergraduate ecotoxicology	☆		Problem-based Learning, Community of Inquiry, distance and blended education	MM (archived online discussion posts, course marks)

(continued)

Table 1.1 (continued)

Chapter	Geographic setting[a]	Grade level	Focus		Topic	Method[b] (Data source)
			Student	Teacher		
10	QU	Grades 7–11		☆	Museums, social learning theory	Museums, social learning theory
11	NL	Grade 10		☆	Place-conscious pedagogy, professional identity	Place-conscious pedagogy, professional identity
12	QU, NU	13–14 years/ youth	☆		Informal science, Indigenous epistemologies, water stewardship	Informal science, Indigenous epistemologies, water stewardship

[a] AB = Alberta, BC = British Columbia, NB = New Brunswick, NL = Newfoundland and Labrador, NS = Nova Scotia, NU = Nunavut, QU = Québec, SK = Saskatchewan
[b] MM = mixed methods, QUAL = qualitative

issues; assessment practices; locally developed courses; instructional approaches; environmental considerations (Tippett & Milford, 2019).

The master list was modified to allow standardization of language (e.g., evaluation and assessment became assessment), separation (e.g., goals and policy, curriculum, evaluation, and assessment was split into goals, policy, and curriculum and assessment), combination (e.g., professional development was merged with teacher education), or elimination (e.g., low priority of science) and then alphabetized. Once we had our master list, we populated Table 1.2 with the three sources and the topics of our elementary and secondary and tertiary research books to facilitate analysis and provide readers with an overview of the topics, trends, and themes.

The collected chapters showcase a selection of contemporary issues in Canadian science education research. This selection aligns with international research trends—although not including educational technology or philosophy, history, and nature of science—while highlighting several unique national concerns that are consistently evident across the top ten SERG topics, science education themes, and elementary, secondary, and tertiary research topics (e.g., environmental issues, Indigenous perspectives). We were surprised that STEM or multidisciplinary approaches did not appear in the collection. We wonder if this omission might be a reflection of the interests of the authors who contributed, or the tendency of upper-level science education to be siloed by teachers and curriculum documents, or both.

Analysis of these 11 chapters through the lens of the consistencies, commonalities, and distinctions that emerged in *Science Education in Canada* (Tippett & Milford, 2019) revealed more overlap in this secondary book than was evident in the elementary research book (Tippett & Milford, 2023a). Of the two consistencies—a low priority of science and influences on science curricula—one was represented, with

Table 1.2 International and national research topics, trends, and themes

Master list of topics	International trends	Top ten SERG topics	Science education themes	Elementary research topics	Secondary & tertiary research topics
Assessment	☆		☆		☆
Culture, society, and gender	☆	☆	☆		☆
Educational technology	☆				
Environmental issues		☆	☆	☆	☆
Goals, policy, and curriculum	☆		☆		☆
Identity and self-efficacy		☆			☆
Indigenous perspectives			☆	☆	☆
Informal learning	☆				☆
Inquiry and other skills		☆		☆	
Learning (conceptions)	☆	☆			☆
Learning (context)	☆			☆	☆
Philosophy, history, and NOS	☆				
STEM and engineering		☆		☆	
Student-focused		☆		☆	
Teacher education and professional development	☆	☆	☆	☆	☆
Teaching	☆	☆	☆	☆	☆

Note NOS = nature of science; SERG = science education research group; STEM = science, technology, engineering, and mathematics

chapters in this book addressing influences on new curriculum at the policy and classroom levels. The four commonalities—Indigenous perspectives, inadequate science teacher education, language issues, and assessment practices—were each represented in at least one chapter. The three distinctions—locally developed courses, instructional approaches, and environmental considerations—were also clear in the chapters exploring place-conscious learning, adapted primary literature, problem-based learning, ocean literacy, and relational approaches.

1.5 Secondary Science Education—What Next?

Our exploration of Canadian secondary science education reveals a critical inflection point. The traditional disciplinary boundaries of science may no longer be tenable in an increasingly complex, interconnected world. This book serves not as a definitive statement, but as a strategic provocation—challenging researchers, educators, and policymakers to reimagine science education as a dynamic, integrative domain with a myriad of possibilities to build upon our international reputation for high-quality science instruction and achievement. By mapping the current landscape, we are laying the foundation for systematic, evidence-driven transformation of how science can be conceptualized, taught, and understood.

We return to our recommendations for changes in policy and practice that would address pressing issues in science education identified in *Science Education in Canada* (Tippett & Milford, 2019):

- updating the Common Framework
- reconceptualizing scientific literacy
- incorporating Indigenous perspectives
- shifting to interdisciplinary models such as STEM, and
- improving science teacher education.

1.5.1 Common Framework

The *Common Framework* (CMEC, 1997) is outdated and desperately in need of a refresh. Unchanged for more than 25 years, it fails to address pressing national, provincial and territorial priorities and rapidly evolving challenges that our students face, such as climate change and artificial intelligence. Moreover, its narrow focus on content knowledge needs to expand and encompass the crucial competencies that students require to navigate today's complex world. These competencies are multi-dimensional capacities that allow individuals to explore issues, understand different perspectives, engage positively with others, and take responsibility for personal and societal well-being (OECD, 2016) and can include critical thinking, problem solving, creativity, innovation, collaboration, and communication. While basic understanding of scientific concepts and processes remains important, the goal of science education must expand beyond knowledge acquisition to include the development of students' capacity for critical evaluation of scientific claims, particularly in an era of widespread misinformation and complex socioscientific issues.

1.5.2 Scientific Literacy

A move to include competencies as a goal of science education would require an updated conceptualization of scientific literacy. In Tippett and Milford (2019), we proposed a multidimensional model that included fundamental (competencies), disciplinary (knowledge), and applied (action) dimensions. Each chapter in this volume clearly illustrates one or more of these dimensions. For example, Raisinghani (Chapter 3) describes the issue of language competency when teaching science to English Language Learners, Vergis (Chapter 8) examines procedural knowledge in university chemistry courses, and Rahm (Chapter 2) presents a relational approach to taking action on the issue of water quality.

1.5.3 Indigenous Perspectives

The relational approach sits at the heart of Indigenous epistemologies and is evidence of the need to incorporate Indigenous perspectives as essential to a full understanding of science. Unfortunately, most Canadian curriculum documents still present Indigenous knowledge as ancillary rather than as essential to scientific understanding (Tippett & Milford, 2019). Meaningful changes that address the calls of the Truth and Reconciliation Commission of Canada (TRC, 2015) requires correcting this supplementary position through actions such as creating partnerships with Indigenous Elders and adopting culturally responsive approaches to science. The goal is a holistic, culturally rich understanding of science and science practices that transcends traditional Western frameworks.

1.5.4 Stem

Although secondary and tertiary science education often positions disciplinary knowledge as separate and distinct discipline-specific subjects, adopting a multidisciplinary approach such as STEM can allow students to integrate procedural and conceptual understanding across fields, fostering innovative solutions to complex real-world problems. STEM education often involves the adoption of problem-based approaches that can encourage critical thinking across multiple disciplines. However, teachers can face multiple challenges when implementing these approaches, such as lack of resources, content knowledge, and time (Chai, 2019).

1.5.5 Improvements in Teacher Education

High-quality secondary and tertiary science education demands innovative, comprehensive teaching approaches that embrace a contemporary understanding of scientific literacy, Indigenous perspectives, and multidisciplinary approaches; teacher education programs must adapt to ensure that novice teachers can meet these demands, breaking away from traditional didactic approaches. Secondary teacher education programs in Canada typically require deep content knowledge for admission and then provide broad pedagogical knowledge during coursework and field experiences (Tippett & Milford, 2017). However, these programs sometimes fall short in the area of pedagogical content knowledge (PCK), the knowledge of "ways of representing and formulating the subject that make it comprehensible to others" (Shulman, 1986, p. 9). While addressing basic pedagogy, teacher education programs often fail to equip preservice teachers with essential PCK needed to promote modern conceptions of scientific literacy, plan and implement multidisciplinary lessons, and integrate Indigenous knowledge. Research suggests that there is a correlation between the number of PCK courses and the development of preservice teachers' PCK (Schiering et al., 2022). Therefore, additional coursework centered on components of PCK could be bolstered with extended science field experiences and enhanced by collaboration with local Indigenous Elders.

1.6 How to Read This Book

Each chapter in this book offers valuable insights for Canadian and international scholars, describing features of a science education system that supports high student achievement on international assessments such as PISA. For example, inquiry-based and problem-based approaches appear to offer promising pathways towards our proposed three-dimensional model of scientific literacy. Several chapters demonstrate how authentic science experiences can be meaningfully integrated into classroom instruction (e.g., Wells & Ricketts, Chapter 5) and informal educational experiences (e.g., Rahm, Chapter 12) in diverse classrooms such as that described by Raisinghani (Chapter 4).

The chapters are ordered along a broad spectrum ranging from policy through curriculum to classroom teaching. Within that spectrum, the focus shifts from teachers to teaching approaches to student learning. However, there is no required order in which to read the chapters, and we invite readers to select an approach that meets their needs. Chapters might be accessed according to topic or because of their focus on teachers or students. Chapters can be read on their own as individual works or may be read as a part of the larger collection. The book itself can be read as a standalone or in conjunction with *Exploring Elementary Science Teaching and Learning in Canada* (Tippett & Milford, 2023a) for a more comprehensive picture of science education research in Kindergarten through Grade 12 and beyond.

Readers interested in learning more about the contexts that inform the research in our elementary and secondary/tertiary level books are encouraged to refer to *Science Education in Canada* (Tippett & Milford, 2019). All three books and their content are intended to serve as provocations that encourage deeper exploration of micro-level aspects of secondary or tertiary science education, enable the consideration of meso-level factors, and support macro-level comparisons across educational systems internationally.

References

Chai, C. S. (2019). Teacher professional development for science, technology, engineering and mathematics (STEM) education: A review from the perspectives of technological pedagogical content (TPACK). *Asia-Pacific Education Researcher, 28*(1), 5–13. https://doi.org/10.1007/s40299-018-0400-7

Council of Ministers of Education, Canada. (n.d.-a). *About us*. https://www.cmec.ca/11/About_Us.html

Council of Ministers of Education, Canada. (n.d.-b). *Over 50 years of Pan-Canadian leadership in education*. https://www.cmec.ca/299/education-in-canada-an-overview/index.html

Council of Ministers of Education, Canada. (1997). *Common framework of science learning outcomes, K to 12: Pan-Canadian protocol for collaboration on school curriculum for use by curriculum developers*. https://archive.org/details/commonframework00coun

Crocker, R. K., & Dibbon, D. C. (2008). *Teacher education in Canada: A baseline study*. Society for the Advancement of Excellence in Education.

Dopfer, K., Foster, J., & Potts, J. (2004). Micro-meso-macro. *Journal of Evolutionary Economics, 14*(3), 263–279. https://doi.org/10.1007/s00191-004-0193-0

Kim, M. (2017). Indigenous knowledge in Canadian science curricula: Cases from Western Canada. *Cultural Studies of Science Education, 12*(3), 605–613. https://doi.org/10.1007/S11422-016-9759-Z

Lin, T.-J., Lin, T.-C., Potvin, P., & Tsai, C.-C. (2018). Research trends in science education from 2013 to 2017: A systematic content analysis of publications in selected journals. *International Journal of Science Education, 41*(3), 367–387. https://doi.org/10.1080/09500693.2018.1550274

MacDonald, R. J. (2010). Bridging the theory-practice divide: Teaching science methods off campus. In T. Falkenberg & H. Smits (Eds.), *Field experiences in the context of reform of Canadian teacher education programs* (Vol. 1, pp. 261–273). Faculty of Education of the University of Manitoba.

Maeng, J. L., & Bell, R. L. (2015). Differentiating science instruction: Secondary science teachers' practices. *International Journal of Science Education, 37*(13), 2065–2090. https://doi.org/10.1080/09500693.2015.1064553

Milford, T. M., & Tippett, C. D. (2016). Canada: Teaching science in Canada. In B. Vlaardingerbroek & N. Taylor (Eds.), *Teacher quality in upper secondary science education* (pp. 43–58). Palgrave Macmillan.

Milford, T. M., & Tippett, C. D. (2019). Introduction: Setting the scene for a meso-level analysis of Canadian science education. In C. D. Tippett & T. M. Milford (Eds.), *Science education in Canada: Consistencies, commonalities, and distinctions* (pp. 1–12). Springer.

Olson, J. K., Tippett, C. D., Milford, T. M., Ohana, C., & Clough, M. (2015). Teacher preparation in a North American context. *Journal of Science Teacher Education, 26*(1), 7–28. https://doi.org/10.1007/s10972-014-9417-9

Organization for Economic Cooperation and Development (OECD). (2016). *PISA. Preparing our youth for an inclusive and sustainable world: The OECD PISA global competence framework.* https://www.oecd.org/education/Global-competency-for-an-inclusive-world.pdf
Organisation for Economic Cooperation and Development. (2018). *PISA 2018 results (Vol. I): What students know and can do.* OECD Publishing. https://doi.org/10.1787/5f07c754-en
Organisation for Economic Cooperation and Development. (2023). *PISA 2022 results (Vol. I): The state of learning and equity in education.* OECD Publishing. https://doi.org/10.1787/53f23881-en
Schiering, D., Sorge, S., Keller, M. M., & Neumann, K. (2022). A proficiency model for pre-service physics teachers' pedagogical content knowledge (PCK)—What constitutes high-level PCK? *Journal of Research in Science Teaching, 60*(1), 136–163. https://doi.org/10.1002/tea.21793
Shulman, L. S. (1986). Those who understand: Knowledge growth in teaching. *Educational Researcher, 15*(4), 4–14. https://doi.org/10.3102/0013189X015002004
Serpa, S., & Ferreira, C. M. (2019). Micro, meso and macro levels of social analysis. *International Journal of Social Science Studies, 7*(3). https://doi.org/10.11114/ijsss.v7i3.4223
Statistics Canada. (2024a). *List of Canadian universities.* https://www23.statcan.gc.ca/imdb/p3VD.pl?Function=getVD&TVD=1269102
Statistics Canada. (2024b). *Postsecondary enrolments, by field of study, registration status, program type, credential type and gender.* https://doi.org/10.25318/3710001101-eng
Tippett, C. D., & Milford, T. M. (2017). Canada: An overview of secondary science teacher education programs. In J. Pedersen, T. Isozaki, & T. Hirano (Eds.), *Model science teacher preparation programs: An international comparison of what works* (pp. 163–184). Information Age Publishing.
Tippett, C. D., & Milford, T. M. (Eds.). (2019). *Science education in Canada: Consistencies, commonalities, and distinctions.* Springer. https://doi.org/10.1007/978-3-030-06191-3
Tippett, C. D., & Milford, T. M. (Eds.). (2023a). *Exploring elementary science teaching and learning in Canada.* Springer. https://doi.org/10.1007/978-3-031-23936-6
Tippett, C. D., & Milford, T. M. (2023b). Providing a space for Canadian science education research. In C. D. Tippett & T. M. Milford (Eds.), *Exploring elementary science teaching and learning in Canada* (pp. 1–8). Springer. https://doi.org/10.1007/978-3-031-23936-6_1
Tippett, C. D., Milford, T. M., & Yore, L. D. (2019). Epilogue: The current context of Canadian science education and issues for further consideration. In C. D. Tippett & T. M. Milford (Eds.), *Science education in Canada: Consistencies, commonalities, and distinctions* (pp. 311–337). Springer. https://doi.org/10.1007/978-3-030-06191-3_15
Tippett, C. D., Pegg, J. M., Wiseman, D., & Milford, T. M. (2017, August 21–25). *Canadian science education research: A comparison across dissemination outlets* [Paper presentation]. European Science Education Research Association, Dublin, Ireland.
Truth and Reconciliation Commission of Canada. (2015). *Honouring the truth, reconciling for the future: Summary of the final report of the Truth and Reconciliation Commission of Canada.* https://ehprnh2mwo3.exactdn.com/wp-content/uploads/2021/01/Executive_Summary_English_Web.pdf
Tsai, C.-C., & Wen, M. L. (2005). Research and trends in science education from 1998 to 2002: A content analysis of publications in selected journals. *International Journal of Science Education, 27*(1), 3–14. https://doi.org/10.1080/0950069042000243727
Walan, S. (2020). Embracing digital technology in science classrooms—Secondary school teachers' enacted teaching and reflections on practice. *Journal of Science Education & Technology, 29*(3), 431–441. https://doi.org/10.1007/s10956-020-09828-6

Chapter 2
Curriculum Concerns: Hearing the Voices of Saskatchewan Teachers Regarding a New Grade 12 Physics Curriculum Document

Ellen Watson

2.1 Introduction

It is rare for a science teacher to have a career without experiencing curriculum change (Ryder et al., 2014). As one Canadian example of frequent change, the British Columbia science curriculum had no less than six shifts between student-centered (i.e., focused on the needs of individual students) to subject-centered (i.e., focused on teaching a subject) curricula between 1920 and 2014, with three of these shifts occurring after 1983 (Sun et al., 2015). School curriculum changes in Canada are determined and mandated by each province. Historically in Saskatchewan, some teachers and other educational stakeholders have been included in both the writing and revising of curricula, based on the legacy of Henry Janzen, a leader in provincial curriculum development. For example, I was one of four Saskatchewan physics teachers who were part of the working group helping to write the 2016 Physics 30 curriculum document (the document discussed throughout this study). There were also seven physics teachers across Saskatchewan who piloted and reviewed this document along with other stakeholders, such as faculty and outreach coordinators, to help guide the creation of the document.

All Saskatchewan teachers are expected to "implement the provincial curriculum conscientiously and diligently" (Saskatchewan Teachers' Federation, 2023, p. 42). The Saskatchewan Ministry of Education (SME) has been conscientious in hearing those teachers selected for integral roles during the curriculum-writing process; however, after curriculum documents are fully released, most teachers are not asked for feedback on those documents (Saskatchewan Teachers' Federation, 2018,

E. Watson (✉)
Brandon University, Brandon, MB, Canada
e-mail: watsone@brandonu.ca

C. D. Tippett and T. M. Milford (eds.), *Seeing Science Through the Eyes of Canadian Teachers and Learners*, Contemporary Trends and Issues in Science Education 54,
https://doi.org/10.1007/978-3-031-88785-7_2

2019). Teachers should be encouraged to question, critically examine, and provide their opinion on initiatives mandated by the educational system (Goldstein, 2014; Watson & Rose, 2019); hence, this study sought to provide teachers with that opportunity.

Used to explore teachers' perceptions of educational change, teacher concern research became part of the educational research landscape with the work of Frances Fuller. *Teacher concerns* are defined as questions, worries, resistances, uncertainties, feelings, and anxieties expressed because of educational changes (Hall & Hord, 2020; Lau & Jong, 2023; van den Berg et al., 2000; Yan & Deng, 2019) and are often classified as responses related to oneself, one's tasks, or the impact on one's students and community (Fuller, 1969). Teacher concerns regarding implementation of curricula have been studied in high school sciences abroad (Borgerding et al., 2013; Gabby et al., 2017; Geng et al., 2019; Gudyanga & Jita, 2018; Ohlemann et al. (2023); Limiansi et al., 2023; Ryder et al., 2014), but studies analyzing teacher concerns regarding secondary science curriculum in a Canadian context are lacking. In addition, the literature investigating teacher concerns specific to physics education is meager as most studies have focused on science education in general. In response to these research gaps and in the study described here, I sought to understand teacher concerns regarding the adoption of a senior physics curriculum document by asking, "What were the concerns of Grade 12 physics teachers in Saskatchewan regarding the 2016 curriculum document?".

2.2 Changes to the Grade 12 Physics Curriculum Document in Saskatchewan

In 1984 the *Directions* report, released by the Saskatchewan Department of Education, promoted the rethinking of public education to ensure students would be prepared for the twenty-first century (Robinson, 2006). Following that report, the Physics 30 curriculum document was created by the Curriculum and Instruction Review Committee and released in 1992. Beginning in 2005, SME began revising its K–12 science curriculum documents after the *Common Framework of Science Learning Outcomes: Pan-Canadian Protocol for Collaboration on School Curriculum, K–12* (commonly referred to as the Pan-Canadian Framework; CMEC, 1997).

The Grade 12 sciences (Physics 30, Chemistry 30, & Biology 30) were the last to be released in the fall of 2016. A noticeable change to align the new curriculum documents with the Pan-Canadian Framework was the use of mandatory, overarching outcomes supported by several optional indicators (SME, 2010). The 1992 Physics 30 curriculum required 244 learning objectives to be taught in four core units, along with topics and learning objectives from at least one of four optional units (Saskatchewan Department of Education, 1992). The units and topics from the 1992 Physics 30 curriculum document are shown in Table 2.1.

Table 2.1 Summary of the units and topics for the 1992 Physics 30 curriculum document (SME, 1992)

Unit	Topics
Core Unit I: Kinematics and Dynamics	A. Understanding Motion B. Vector and Scalar Quantities C. Distance and Displacement D. Speed and Velocity E. Acceleration F. Newton's Laws of Motion
Core Unit II: Mechanical Energy	A. Work B. Power C. Kinetic Energy D. Gravitational Potential Energy
Core Unit III: Electricity	A. Applications B. Current and Potential Difference C. Electric Circuits D. Electric Power and Energy
Core Unit IV: Nuclear Physics	A. Atomic Theory B. Nuclear Fission C. Nuclear Reactors
Optional Unit V: Applications of Kinematics and Dynamics	A. Momentum B. Frictional Forces C. Projectile Motion D. Uniform Circular Motion E. Universal Gravitation
Optional Unit VI: Fluid Mechanics	A. Density B. Pressure C. Pascal's Principle D. Archimedes Principle E. Bernoulli's Principle
Optional Unit VII: Electromagnetism	A. Magnetism B. Electromagnetism C. The Motor Principle D. Electromagnetic Induction
Optional Unit VIII: Atomic Physics	A. Natural Radioactivity B. Half Life and Radioactive Decay C. Nuclear Fusion D. Applications E. Contemporary Physics

The 2016 Physics 30 curriculum required that students achieve nine overarching outcomes across five units (SME, 2016): student-directed study (one outcome), modern physics (two outcomes), forces and motion (two outcomes), conservation laws (two outcomes), and fields (two outcomes). Table 2.2 shows one of the outcomes from the 2016 curriculum and its associated indicators. Indicators represent examples that could support an understanding of an outcome, are not mandatory, and "describe ways that students might demonstrate achievement of an outcome and clarify the breadth and depth of each outcome" (Molnar et al., 2019, p. 71).

All the updated senior science curricula contained a unit on student-directed study; the 2016 Physics 30 curriculum also outlined four other mandatory units: modern physics, forces and motion, conservation laws, and fields. Modern physics and fields were new units although a few of the indicators under the overarching outcomes were learning objectives included in optional units within the 1992 curriculum document. For other units, topics from the 1992 curriculum document were rearranged and included as indicators under new outcomes in the 2016 document. Many of the topics under the 1992 unit of kinematics and dynamics were migrated into other science courses; some kinematics topics were moved into Science 10, while dynamics topics were included under the forces and motion unit in the 2016 Physics 30 course. Given these substantial changes, my research explored teachers' reactions to this new curriculum document.

2.3 Conceptual Framework

The Stages of Concern Framework (SoC; Hall et al., 1973), an aspect of the Concerns-Based Adoption Model (CBAM, Fuller, 1969), was used to conceptualize teachers' concerns about the 2016 Physics 30 curriculum document. CBAM has been used for over 50 years to measure, describe, explore, and explain teacher experiences during curricular change (Anderson, 1997; Hall & Hord, 2020). CBAM provides a language and organization for those feelings that teachers experience during a change and is based on three main assumptions: (a) change is a process, not an event; (b) change is personal and involves developmental growth; and (c) change can be facilitated (Anderson, 1997; Borgerding et al., 2013; Hall & Hord, 2020). CBAM offers a tested, well-developed theoretical model from which to explore teacher experiences.

Fuller's (1969) original CBAM contained three phases: pre-teaching, early teaching (concern with self), and late teaching (concern with students). Hall et al. (1973) extended these phases of concern into seven progressive stages to create the Stages of Concern (SoC) Framework. The stages, as paraphrased from Anderson (1997), are:

1. *Unconcerned:* The teacher is uninterested in or unaware of the change.
2. *Informational:* The teacher wants to learn more about the change and its implications. The teacher begins to learn about the change.
3. *Personal:* The teacher is anxious about their role in the change and may participate, but they are not overly interested in the innovation.
4. *Management:* The teacher begins to explore the change and concerns focus on logistics.
5. *Consequence*: The teacher concerns shift to focus on the impact the change will have on students in their classroom and how to modify the change to fit personal practice.

Table 2.2 Example outcome and indicators from the Physics 30 curriculum (SME, 2016, p. 31)

Outcome	Indicators
PH30-CO2 Analyze the motion of objects and interactions between objects using momentum concepts, including the law of conservation of momentum	Explore how impulse and momentum concepts apply to motion-related technologies in fields such as sports science, transportation and space science
	Derive the formula for impulse from Newton's second law of motion
	Investigate how changing the net force applied to an object or the amount of time the force is applied affect the momentum of an object, with reference to the formula $\vec{F}\,\Delta t = m\Delta\vec{v}$
	Provide examples that show how momentum is or is not conserved in everyday situations
	Conduct an experiment or simulation, including collecting, analyzing and interpreting data, to determine the extent to which momentum is conserved in elastic and inelastic collisions
	Solve problems using the law of conservation of momentum in one- and two-dimensional interactions (e.g., head-on collisions, glancing collisions, rocket launches and explosions)
	Analyze applications (e.g. neutrino detection, Large Hadron Collider, crash test dummies and personal safety devices) of the law of conservation of momentum

6. *Collaboration:* The teacher begins to show interest in working with others to embrace change and attempt implementation to best improve local context.
7. *Refocusing:* The teacher has deeply considered the innovation and their concerns revolve around modifying innovation to reap even more of the intended benefits.

These seven stages provide a framework with which teachers' concerns can be viewed as they relate to the self (Stages 0, 1, & 2), tasks (Stage 3), and others (Stages 4, 5, & 6).

2.4 Method

Teachers for this study were recruited through multiple approaches: (1) snowball sampling through personal contacts, (2) all school divisions across Saskatchewan were invited to share information about this study with their teachers, and (3) through social media posting. Sixteen Grade 12 physics teachers across Saskatchewan volunteered to discuss their concerns about the 2016 curriculum document. These teachers came from both urban ($n = 9$) and rural ($n = 7$) schools; urban schools were defined as located in a city or within commuting distance of a city with a population greater than 10,000. Both female ($n = 5$) and male ($n = 11$) teachers were represented, and teaching experience ranged from 6 to over 20 years. Ten teachers were accredited to teach physics in the province, meaning they had taught for at least 2 years and taken a minimum of seven university courses in physics (or related subjects) along with one secondary-level curriculum science methods courses. In Saskatchewan, accredited teachers can write their own final examination for Grade 12 courses and do not need to participate in the provincially administered examination for that course.

2.4.1 Data Collection

As is often the case in researching teacher concerns (e.g., Abadie & Bista, 2018; Borgerding et al., 2013; Fischer et al., 2019; Gabby et al., 2017; Le Fevre, 2014; Limiansi et al., 2023), I conducted all 16 interviews using semi structured questions to guide interviews. I asked questions about their opinions and concerns about the 2016 Physics 30 curriculum document, such as:

- When you first read the 2016 Physics 30 curriculum document, what did you think?
- Has your reaction changed since you first interacted with the document? If so, how?
- Were you surprised by any of the changes made to the curriculum document? How did this surprise (or lack thereof) impact your interaction with the document?
- What questions do you have about teaching using the 2016 Physics 30 curriculum document or the document itself?

- What do you like about the 2016 Physics 30 curriculum document?
- What (if any) criticisms do you have about the 2016 Physics 30 curriculum document?
- How do you think the 2016 Physics 30 curriculum document has (or will) influence your physics teaching?

Following each interview, I wrote a short summary of the discussion and my interpretations, maintaining a list of themes and interesting questions that surfaced during the process. When a new theme or question arose from an interview, I conducted a follow-up interview with previously interviewed teachers to seek their opinions and perceptions on these themes or questions and integrated this new theme/question in future interviews.

2.4.2 *Data Analysis*

I transcribed the interviews and coded them using thematic analysis, a process where the researcher identifies, analyzes, and reports patterns (or themes) viewed within data (Braun & Clarke, 2006, 2021). Thematic analysis involves six phases:

- familiarize yourself with the data
- generate codes (unless using an a priori framework) and code the data
- search for themes
- review themes
- define and name the themes
- select exemplars to represent the theme and report.

Thematic analysis can be either deductive (i.e., using pre-existing frameworks as initial codes), inductive (i.e., looking for themes generated from data), or, as in this study, both. Initially, I coded statements in the transcripts using themes that I documented during the interviews (e.g., worried about teaching new content, preparing students for a provincial examination, a lack of materials). During coding, I revisited themes, occasionally collapsing some together, diffracting some into multiple themes, or, in one case, removing a theme altogether as it had insufficient supporting evidence.

2.4.3 *Member Checking*

To verify the representations of reality that I constructed, I conducted member checks (Guba & Lincoln, 1989). I sent the transcripts, along with a list summarizing my interpreted teacher concerns, to each participant asking for feedback on my interpretations. None of the participants requested changes to their transcripts nor to the teacher concerns that I identified.

2.5 Results

I identified ten concern themes, each describing a different concern held by multiple teachers. Next, I connected each theme to a stage of concern (Hall & Hord, 2020), as shown in Table 2.3. Finally, I selected exemplar statements representing each theme and wrote a description of the theme using pseudonyms.

The participating Physics 30 teachers were eager to discuss their concerns, worries, questions, and anxieties regarding the 2016 Physics 30 curriculum document during their interview. Next, I briefly discuss each concern identified in this section, which is organized by the Stages of Concern Framework.

2.5.1 Stage 2: Personal Concerns

In this study, teachers' personal concerns largely aligned with two themes: discomfort with content and ensuring success on the provincial exams. Teachers in this study were concerned about how these two areas might impact their abilities to effectively teach Physics 30. Scholars often characterize people in Stage 2 as being largely uninterested in participating in a particular innovation (Anderson, 1997; Hall & Hord, 2020). However, because teaching this curriculum was not a choice—all teachers were required to implement it—disinterest was likely not the case in this study.

2.5.1.1 Teacher Discomfort with Content

Nine of the 16 teachers interviewed expressed some discomfort with teaching the two new units in the 2016 curriculum (i.e., fields and modern physics). Teachers were concerned they did not know this content well enough to teach. As an example, Jens,

Table 2.3 Concern themes as aligned with the stages of concern

Concern Themes	Stage of Concern
1. Teacher discomfort with content 2. Teacher effectiveness in ensuring success on the provincial examination	Stage 2: Personal
3. Lack of instructions and resources 4. Lack of time to cover content	Stage 3: Management
5. Preparation of new/non-science teachers 6. Student lack of prerequisite knowledge 7. Content is too abstract 8. Course variation across the province	Stage 4: Consequence
9. Need to connect with other physics teachers 10. Resistance among other teachers	Stage 5: Collaboration

Note Only those stages of concern with themes present have been included

a physics teacher of almost 10 years who held degrees in chemistry and education, said, "I was definitely apprehensive about teaching the modern physics because, with a limited physics background, I didn't really know a ton about it." Similarly, Harley, an accredited physics teacher for over 20 years, said she still was concerned about teaching fields because she didn't "know a lot about it and there aren't very many resources about it." Kye, an accredited physics teacher with approximately 15 years of experience, summarized the teachers' sentiments nicely: "I have to make sure my understanding is good and then [I can] bring it from where it is in my head to [my students]." Teachers concerned with teaching unfamiliar content were worried about ensuring they fully understood these ideas before they could be effective in teaching them to their students. No teacher expressed concerns about teaching those topics that appeared in the 1992 curriculum document.

2.5.1.2 Teacher Effectiveness in Ensuring Success on the Provincial Examination

The 2016 document was designed to encourage teacher—and student—exploration with required outcomes supported by optional indicators. This format was intended to allow Physics 30 teachers more autonomy in designing the course. For accredited teachers, this meant more freedom in how and what they wanted to teach. Harley "felt guilty" because, as an accredited teacher, she had the luxury of spending less time on some outcomes in order to go deeper into other outcomes and non-accredited teachers had to "cover everything." The 2016 curriculum document design was intended to increase teachers' freedom when designing their Physics 30 course, yet non-accredited teachers found the outcome-indicator structure frustrating.

Three of the six non-accredited teachers (whose students had to write the Physics 30 provincial examination) raised concerns regarding their effectiveness in preparing students to write the provincial examination. Most often, these concerns rose from tension between teaching the curriculum document and preparing students to write a provincially administered examination. Franz felt "the whole idea of having outcomes and indicators is really useless to me because of the provincial exam[ination]." He found the open format frustrating because the provincial examination "restricts me.… Ultimately, I have to teach the indicators because that's what's going to be evaluated."

Non-accredited teachers struggled to discern what indicators would be assessed by the provincial examination. For example, Brad asked, "How are we going to know—if we just have indicators—what [students] need to know for their final exam since teachers are supposed to be [teaching] just some indicators." This was perceived as being different than the 1992 curriculum document that, according to Brad, had "things on a checklist, … [so I could] tick them off and make sure I did all of them."

2.5.2 *Stage 3: Management Concerns*

The management concerns of teachers in this study were primarily connected to either a lack of instructions and resources or a feeling that there was too much content to cover in one course. Teachers were unsure about the expectations of the curriculum document and wanted more direction on how to dedicate their instructional time. These concerns were primarily focused on the logistics of the implementation of the 2016 Physics 30 curriculum document.

2.5.2.1 Lack of Instructions and Resources

Eight teachers expressed concern over a lack of instructions on how to teach the content in the 2016 curriculum document and a lack of resources with which to teach this content. Two teachers, Jens and Gru, questioned the lack of direction on how much time should be spent on each outcome. Gru commented, "I missed the recommended time to spend on stuff." The 1992 curriculum document indicated how many hours should be dedicated to each unit (e.g., 8–10 h for the work and energy unit), whereas the 2016 document contained no time recommendations. "[There are topics in this curriculum that] you can design an entire course on … so it's kind of hard to know how to allocate your time," noted Jens. Gru "like[d] that [SME] treated [teachers] as professionals, able to make their own decisions but, … having some criteria to make those decisions upon is useful." Both Jens and Gru were concerned about the lack of guidance given on time allocation to units with the 2016 curriculum document.

Teachers expressed concerns regarding how they might enact the 2016 curriculum and, in particular, about a lack of available resources. These concerns ranged from Nadia requesting a specific layout of what to teach: "whenever they bring in a new curriculum, they should give [teachers] a binder and say this is basically it" to Olivia describing the problem of needing too many resources: "I used lots of different textbooks … to cover every topic that's covered in the curriculum because there's no perfect textbook." Egon shared that "there is less supporting material [with this 2016 curriculum and this was] especially [true] for that electricity and fields unit." Both Harley and Chaz discussed available resources, but most interviewed teachers were unaware of them. Half of the interviewed teachers expressed frustration with a lack of identified resources supporting the 2016 Physics 30 curriculum, especially with those units new to the course (i.e., modern physics and fields).

2.5.2.2 Lack of Time to Cover Content

Five teachers (3 accredited and 2 non-accredited) felt the 2016 Physics 30 curriculum contained too much content to cover in one semester. As Alan commented, "So, they took out 5 days of my lessons and added on 12. Great. How am I going to do this?" One potential reason for this concern, as expressed by Olivia, was having a lot of indicators for each learning outcome and that some of these indicators, such as velocity and vectors, required a substantial amount of time to teach. As Olivia was not accredited in physics, she felt responsible for covering all the potential indicators that could be assessed on the provincial examination. The 2016 curriculum was structured in such a way that teachers needed to cover only nine overarching outcomes; however, the perceived amount of content associated with some of the indicators made covering the entire document difficult in the eyes of these five teachers.

For Chaz, an accredited physics teacher, there was too much content to teach the entirety of the 2016 Physics 30 course because the new document lacked "fundamental content" that needed to be taught in addition to the expected indicators. Chad explained, "When I say that I think there's too much stuff in [this curriculum], it's because I still teach the fundamental core stuff that was cut out." Specifically, Chaz mentioned the need to teach vectors and review content that was now part of Science 10, which he felt was often poorly taught. Chaz indicated that he had asked a Saskatchewan curriculum consultant why kinematics and vectors had been removed from the 2016 curriculum document:

> [the consultant] said, 'we teach it in Grade 10.' I was like, 'Really? So, you expect kids to remember what they learned in Grade 10 all the way in Grade 12?' Grade 10 is a write-off year—and this is coming from a front-line person.

Again, teachers saw the 2016 Physics 30 curriculum document as larger than intended because of implied content.

In contradiction, two teachers claimed that it was easier to cover the content in the new document than in the previous Physics 30 curriculum. Ian, a long-time accredited teacher, felt that the science consultant "did unburden [this new] curriculum. I had never, in my entire time with the previous curriculum, finished it. Never." To Ian, the fact that he was able to finish the Physics 30 course indicated that the 2016 document was more manageable than its predecessor. Egon, a non-accredited teacher, also felt less pressed for time; he "moved at a brisk pace but was still able to finish it with time for review…. In terms of actual curriculum, [he] wouldn't say it is too much—it's sufficient." The amount of content in the 2016 curriculum document overwhelmed some teachers, but others found they had adequate time to teach the complete course.

2.5.3 *Stage 4: Consequence Concerns*

Many of the concerns expressed by teachers in this study focused on modifying their teaching practice to meet the parameters of the 2016 curriculum document. These concerns were characterized as focusing on the impact of teaching changes on student learning. Teachers' concerns in this stage, for this study, can be described by the themes preparing new (or non-) science teachers, concerns about students' lacking pre-requisite knowledge, teaching increasingly abstract content, and course variation across the province.

2.5.3.1 Preparation of New/Non-Science Teachers

Seven teachers were concerned about the success that new teachers, or those without physics as a subject area, might experience when teaching the 2016 Physics 30 curriculum. According to Gru, "people who need the curriculum [document] the most are usually those that are least qualified to teach that subject; I would say that [this 2016 document] might not be as useful for them as the 1992 document." Clarifying further, Franz shared that,

> it was really easy to know what to teach [with the old curriculum] because it was, like, this equation, this equation, this equation, and focus on this concept, and this concept, and this concept and they even gave you examples; [it] led you. So, as a beginning teacher it was such a nice curriculum because you could just really stick to it.

The 1992 curriculum was perceived to be easier for new teachers to enact because it included a brief explanation of the concepts and key ideas in each topic to be covered, numbered learning outcomes the teacher was expected to cover, teaching suggestions, activities, and possible demonstrations. Franz, Gru, and others believed the 2016 curriculum could be more difficult for new or non-science teachers since the document did not provide detailed explanations about content or teaching strategies.

Further, some teachers were concerned about assumptions made by those mandating and assessing the 2016 curriculum document. As one example, one year Nadia was part of the writing group for a provincial exam in another subject. She described a colleague who wanted to include questions on a provincial examination that were not directly connected to the curriculum; to defend his request, he said "but everybody knows that when you teach this, you use this idea." Nadia and her colleagues who were developing the exam that year refused to use these kinds of questions on the grounds that those teachers preparing students for provincial examinations had, typically, less formal preparation in the subject of physics. To Nadia, teachers knowledgeable in physics would be likely to teach this assumed content because everybody experienced in a subject knew to teach these concepts, but the 2016 curriculum document did not spell out these assumptions.

The assumption that people "should know" is problematic for those new to teaching and those not educated in physics. Therefore, some teachers perceived the 2016 Physics 30 curriculum document as being written for those well-versed in physics. Jens summarized these sentiments well when he said, "as long as you have the necessary background [in physics], I don't think it's any more difficult to implement [this document]", but what about those teachers lacking this background?

2.5.3.2 Student Lack of Prerequisite Knowledge

Teachers were concerned about the preparation of students entering Physics 30 courses and attributed a lack of preparation to the current structuring of physics content in the high school science curricula. In Grade 10 Science, students encounter one physics unit on motion. Should students decide to pursue either physics or chemistry, they enroll in one prerequisite course, Physical Science 20, that discusses the physics topics of heat and waves. There is no mathematics prerequisite for Physics 30. Teachers were concerned that students' preparation for Physics 30 was not sufficient, leading to negative consequences such as time lost to review and weaker physics students.

Teachers were concerned about students encountering the basics of motion in Science 10. Denise explained, "There's 12 months, sometimes 18, from talking about kinematics [in Grade 10] to Physics 30, so now we're talking about dynamics and they don't remember it. So, I struggle with how much kinematics was taken out of Grade 12." It was difficult to extend those concepts in Physics 30 with prerequisite concepts taught in Grade 10 Science since 1 or 2 years elapsed before the concepts were revisited, that is, if they were even taught those concepts. Chaz stated that "in Grade 10 most of the physics doesn't get taught. Right? If those teachers run out of time, physics is not their background, so they don't [give it enough class time]."

To mitigate these concerns, Physics 30 teachers reviewed the concepts they perceived to be prerequisite. For the most part, these now-reviewed concepts were included in the old Physics 30 curriculum document but are now assumed to be previously learned. For teachers, Alan said, "the flow from Grade 10 to Grade 12 in physics [wasn't] cohesive so students just [weren't] prepared to go into Physics 30 from Grade 10 Science."

Specifically, teachers expressed the belief that Physical Science 20 did not prepare students for Physics 30. Gru wished that Physical Science 20 "wasn't a prerequisite because I don't see that there is actually prerequisite knowledge in it." Students struggled to make connections between these courses and, as a result, were not well prepared for Physics 30. This issue also existed with the former curriculum because Physics 20 content was not easily connected to Physics 30. Yet, teachers felt that this disconnect in material was compounded by Physical Sciences 20, consisting of a combination of physics and chemistry topics, being often taught by non-physics people. Marcos explained, "Schools have teachers who aren't physics teachers teaching physical science…. We can't delve into the two sciences unless you have people who know both subjects doing it." Many teachers questioned the capability of one science course (i.e., Physical Science 20) to prepare students for both Chemistry 30 and Physics 30 because physics often lost out in Physical Science 20. Teachers were concerned that the structure of the new Grades 10–12 science curriculum produced ill-prepared Physics 30 students.

2.5.3.3 Content is Too Abstract

Some teachers were concerned about the shift from what they called "concrete" topics to "abstract" ideas. A commonly cited example was the change from learning circuit analysis to learning about electric and magnetic fields, which was mentioned by Egon, Jens, Marcos, and Pharris. Egon said,

> From what I've heard, overall people are happy with the [modern physics] but that fields unit, everyone I've talked with seems to miss the old electricity.... I think it's because the old electricity [unit] was more tangible—more, here is something you can do and actually see the light bulb go on. As opposed to things like magnetic fields where, yes, we can see filings moving but we can't measure it as easily.

Marcos noted that "When physics is measurable, labs are easily done and it's easier to teach with activities, which engages students and that's what we want." For participating teachers, some of the new Physics 30 content was less tangible than those topics in the old document. Pharris pointed out that "If you can see things and measure them, then it's just easier to make sense of the physics." Teachers felt measurable and concrete ideas were better suited to Physics 30; Jens explained that "when it's more abstract, I find it a little bit harder for students to grasp the ideas." Teachers found the increase in abstraction with the 2016 curriculum problematic since students struggled to access abstract content, unlike the concrete ideas in the old physics curriculum.

2.5.3.4 Course Variation Across the Province

Participants were concerned that the interpretation required of the 2016 curriculum document, with its outcome and indicator structure, may cultivate many different Physics 30 courses across the province. Gru suggested that "The probability of picking two teachers teaching [a new Physics 30] course [in Saskatchewan] the exact same way is low." Some participants were concerned that other teachers may not teach Physics 30 in a way that they believed physics should be taught. Alan, who liked to challenge students by ensuring that they solved mathematically rigorous problems, was concerned that teachers may not challenge their students enough: "I see an out for really challenging [students] in some topics; I don't act on that interpretation, but I could, and I know some do, and *that* scares me." To Alan, the variation in courses caused by teacher interpretation may result in "less-prepared" Physics 30 graduates.

On the other hand, Harley was concerned about how teachers enacted the 2016 Physics 30 curriculum:

> I worry that, because indicators aren't mandatory, that there's lots of teachers who are teaching [Physics 30] in a teacher-directed way and not inquiry-based. I think it's very possible to teach it that way; you can focus on the indicators that are just the math[ematics].

To Harley, Physics 30 should be inquiry-based with a focus beyond mathematics. However, as accredited teachers have the luxury of selecting from possible indicators, other teachers may not teach the indicators that were important to Harley. The concerns of these two teachers seem at odds with each other: one calls for more mathematical training and the other for more exploration in conceptual physics, yet they are two veins of the same concern. Teachers were concerned about course variation across the province due to interpretation or misinterpretation of what constitutes acceptable evidence for meeting the outcomes of the curriculum.

Finally, some teachers were concerned that students' experience of the 2016 Physics 30 curriculum depended more heavily on their teacher than with the previous document. Franz noted that

> The old curriculum put everybody on a level field and, you know, I think it allowed for weak teachers to compete with good teachers. The new curriculum really depends on the teacher because the way you interpret the curriculum could be so different.

With less formal direction on what to teach, and more maneuverability in specific topics, teachers were concerned that students' experience of the 2016 Physics 30 relied on the teacher's physics expertise. Those teachers who knew physics well could make assumptions unbeknownst to less experienced teachers and include these assumptions in their course design. Students with less experienced physics teachers could be at a disadvantage due to the more open curriculum document requiring implicit knowledge. Course variation, due to interpretation and hidden assumptions, concerned teachers since students across Saskatchewan may have very different experiences with physics.

2.5.4 Stage 5: Collaboration Concerns

Under Stage 5, there were two categories of concerns. Teachers wanted to talk with other teachers about the 2016 curriculum document and they worried about other teachers resisting the adoption of this curriculum.

2.5.4.1 Need for Professional Development

Some physics teachers expressed feelings of isolation. Comments such as Alan's: "I don't have enough colleagues [who teach physics] to validate my assessments so I have to figure it out myself." and Brad's: "I'd like to get together to talk physics with other teachers but don't have anyone here." showcased the seclusion physics teachers felt in both urban (Alan) and rural (Brad) contexts. Teachers wondered how others implemented the 2016 curriculum, but they had no community of peers with whom they could easily connect. As Franz pointed out, they "didn't have access to that type of community." Typically, Saskatchewan schools have one physics teacher; some larger, urban schools have two physics teachers and, very rarely, three or more. Nadia described the situation well:

> When you're a physics teacher, you're kind of on your own. You're the only physics teacher, so you don't get to have a conversation with other people about what they're doing and what they think is important and how they interpret the curriculum and all those kinds of things.

Teachers wanted to talk with other physics teachers about physics and the 2016 curriculum. Franz specified, "What I really need is just chit-chat; just to sit down with somebody or go to their school, you know, take a couple of hours and just go through how they do things." A few teachers appreciated this study for alleviating this isolation. Brad indicated that "it felt good to be able to talk about what physics is about and think about what physics sort of means.... I think about these things, but I don't often have anyone to talk to them about." Teachers were curious about how others taught Physics 30 and wanted to connect with other physics teachers.

2.5.4.2 Resistance Among Other Teachers

Six teachers expressed concerns that their colleagues may resist teaching the 2016 curriculum document as presented; their colleagues might have wanted to continue teaching physics in the same way. Franz commented,

> There's a lot of baggage from the old curriculum that's come into the new one. Teachers still heavily teach kinematics and whatnot with an emphasis on the math[ematics]. I think a lot of teachers still just bust out problems and problem solving because it's easy.

Despite changes in the curriculum document, some were concerned that their peers had continued to teach much of the content the way it was explained in the old Physics 30 document. Both Marcos and Ian, experienced physics teachers, were unsurprised by this perceived resistance. Ian said that the 2016 curriculum "takes away some of [teachers'] favourite stuff and it highlights stuff they're not comfortable with." While concerned their peers might resist this content, Ian and Marcos felt this defiance would lessen with time. Marcos elaborated: "There are old guard out there who have a certain mind frame.... So, you have to wait for the next generation to see change." Neither Marcos nor Ian were surprised by, or too concerned about, the resistance to the 2016 curriculum document as they felt this would lessen as new teachers began to take over the Physics 30 courses.

2.6 Discussion

Teachers' concerns regarding the 2016 Physics 30 curriculum document in Saskatchewan focused on personal effectiveness, managing its implementation, and consequences resulting from the 2016 curriculum. Some teachers expressed a lack of connection with other physics teachers. Overall, four assertions can be considered representative of this study. First, and most emphatically, teachers were concerned about the increased interpretation that the 2016 document required of teachers when compared with the old curriculum document. Second, teachers questioned the sequencing of physics content in the new senior sciences curricula. Third, teachers were concerned about teaching new content while removing previous content from Physics 30. Fourth, physics teachers frequently felt isolated in their communities. These concerns are context-specific but, as I will demonstrate, corroborate other studies investigating concerns regarding science curriculum change.

2.6.1 Increased Interpretation Concerned Teachers

Teachers were concerned about the increased interpretation required to teach the 2016 Physics 30 curriculum document when compared to the old, more prescriptive document. This study highlighted teachers' frustration when teaching from an open curriculum document while simultaneously preparing students for an externally administered examination as other studies also have (Abadie & Bista, 2018; Fischer et al., 2019; Gabby et al., 2017; Ryder et al., 2014; Wallace & Priestly, 2017). Teachers were concerned about the time required to teach the 2016 Physics 30 course. For example, both Olivia (claiming some indicators are too large to be a single indicator) and Chaz (claiming the document is missing fundamental content) spoke to the content as it was interpreted. Indicators were meant to showcase the depth to which teachers should explore each outcome in the Physics 30 curriculum, but teachers in this study also wanted clarification regarding the intended depth of indicators.

Teachers' concerns about covering the content in this curriculum reveal a more pressing question: What is the purpose of a curriculum document for practicing teachers? Is the curriculum document a checklist of content to be covered (as discussed by some participants)? If so, it is not surprising this myth of coverage nags at teachers. The 2016 Physics 30 document provides "the intended learning outcomes that Physics 30 students are expected to achieve in science by the end of the course" (SME, 2016, p. 3), with indicators suggesting how achievement might be demonstrated. However, teachers' concerns about consistent Physics 30 instruction across the province, hidden assumptions, and preparation for the provincial examination point to an understanding of the curriculum document as a checklist of content to be covered.

2.6.2 Course Sequencing Concerns

Participating teachers were concerned the new sequencing of secondary science courses did not adequately prepare students for Physics 30. In the 1992 structure, students could take a course dedicated to physics in Grade 11—Physics 20—but this course was not required as the sole pre-requisite to take Physics 30. Students could meet the pre-requisites for Physics 30 (per the 1992 document) with either Physics 20 or with Science 10 and an algebra-based Grade 11 mathematics course.

Along with the renewal of Saskatchewan science curriculum documents, there was also a significant change to the Grade 11 courses offered. As shown in Table 2.4, there was no longer a course dedicated to physics in Grade 11. Instead, Saskatchewan offers a Physical Science 20 course which covers content traditionally taught in both chemistry and physics. Physical Science 20 is a pre-requisite for Physics 30 in Saskatchewan but can also serve as a pre-requisite for Chemistry and Earth Science 30.

Table 2.4 Science courses and topics in Saskatchewan in grades 10, 11, and 12 (from Molnar et al., 2019, p. 73)

Grade 10	Grade 11	Grade 12
Science 10 • Climate and ecosystem dynamics • Chemical reactions • Force and motion in our world	Environmental science 20 Health science 20 Physical science 20 Computer science 20	Biology 30 Physics 30 Earth science 30 Computer science 30

In this study, teachers voiced concerns about expecting topics such as kinematics to be remembered from Science 10 and, consequently, Physics 30 teachers were unable to cover all the content because course time was needed to review. Teachers highlighted a lack of visible connection between the topics in Physical Sciences 20 and Physics 30. Physics is a logical and hierarchical discipline (Mäntäyla & Nousiainen, 2014; Wheelahan, 2010); thus, it was not surprising that teachers wanted the sequencing of physics in the curriculum to reflect this structure. Akalamkam (2013) found that teachers in India preferred teaching major topics that could be connected to ideas across multiple grades and that this connection helped learners mature in physics. Participating Saskatchewan teachers echoed this finding by asking for clearer connection of the physics content in Science 10, Physical Sciences 20, and Physics 30.

Seven participating teachers shared concerns regarding the Physical Science 20 course where students encounter kinetic molecular theory and properties of waves, but the kinematics behind wave motion is not addressed. As Physical Sciences 20 was often taught by "chemistry people," these teachers were not as well prepared to teach physics as they are to teach the four chemistry-specific outcomes. In the early twentieth century, American educators recommended teaching general sciences courses (combining physics and chemistry), not unlike Physical Sciences 20. However, physicists such as Robert Millikan heavily criticized these courses, claiming the course wasted students' time by teaching superficial understandings (Meltzer & Otero, 2015). Likewise, these participating physics teachers were concerned that the combined structure of Physical Sciences 20 inadequately prepared students for Physics 30.

2.6.3 Concerns About Changed Content

Nine participating teachers were concerned about effectively teaching new content (i.e., modern physics and fields) in the 2016 Physics 30 curriculum. Other studies have shown that when teachers felt inadequately prepared to teach new content, they were uneasy teaching those topics (Borgerding et al., 2013; Limiansi et al., 2023; Lowe & Appleton, 2015; Ramnarian & Fortus, 2013). Considering science, technology, engineering, and mathematics (STEM) education in Hong Kong, Lau and Jong (2023) found that teachers with more content training focused their concerns on others (Stages 4–6). Engaging Saskatchewan teachers with the content new to Physics 30 with the 2016 document could have better prepared teachers to implement the document, moving their concerns up the SoC.

Participating teachers such as Egon and Marcos were concerned about replacing tangible content (e.g., electric circuits) with abstract ideas (e.g., electric fields). Science teachers often see themselves as guardians of content (Mulhall & Gunstone, 2008; Tytler, 2010) and are likely to turn their focus to the familiar when faced with unfamiliar topics (Le Fevre, 2014; Lowe & Appleton, 2015). Pedagogical content knowledge and confidence in a subject area rely on the quality of content knowledge (Kind & Chan, 2017; Wei & Chen, 2019); hence, it is easier for teachers to focus on the areas they know. The "old guard" (as Marcus labelled them) carried with them "baggage" (as Franz put it) from the old curriculum. Physics 30 had been consistent in its format for close to 30 years. After that long, Saskatchewan teachers had likely become comfortable in seeing Physics 30 a certain way; hence, teachers were concerned about content changes, specifically, topics new to the course.

2.6.4 Curiosity About Colleague Uptake

Participating teachers' curiosity regarding how other people taught the 2016 Physics 30 course was related to the isolation they felt. These teachers appreciated the opportunity to discuss physics and the 2016 Physics 30 curriculum that was offered by this study. This finding supports literature claiming physics teachers are often isolated as subject-specialists and can feel disconnected from their colleagues (Kelly & Sheppard, 2010; Nehmeh & Kelly, 2018). Several teachers in this study openly wished they had a community with whom they could discuss teaching physics. Tesfaye and White (2012) found that physics teachers are less likely to connect with science-general professional organizations than teachers of other sciences, and my findings may reflect this lack of connection.

2.7 Concluding Thoughts

Findings from this study highlight the need for ongoing communication with teachers when implementing new senior science curricula. Participating teachers were most emphatic with concerns regarding the management and consequences of the 2016 Physics 30 curriculum document. Saskatchewan science curricula are designed to allow teachers to exercise their professional judgment in deciding how to meet the outcomes of a course. However, many participating teachers struggled with this freedom and felt more specific direction from the curriculum could create a more "even playing field" for teachers independent of their level of experience with physics. This concern was exacerbated by a lack of communication among physics teachers across the province as many see themselves as isolated subject-specialists.

Teachers need more support in learning the content themselves when implementing a new curriculum document with content that differs from previous documents. Participants in this study highlight that this recommendation is particularly important in a subject which has more out-of-field teachers than most, which is the case for physics across North America (Sunal et al., 2019). Since the release of this curriculum document, some resources have been developed to support teaching. However, when curriculum documents are first released in Saskatchewan, it is often the case (as these participants indicate) that there is a lack of supporting material. Teachers' implementation of science curriculum is more impactful when teachers can learn the content and how to teach it using curricular activities (Roehrig, 2023). It can take time to develop these resources, however, it could be beneficial for curriculum developers to engage teachers with the content and curricular resources aimed at content newly introduced to the course as early as possible.

Further, this study suggests that teachers would benefit from additional support and connection when implementing a new curriculum document. Ohlemann et al. (2023) identified that continuous flow both to and between stakeholders during curriculum change can help shift the focus from self- and task- related concerns to concerns focused on others. To facilitate this flow of information, connections can made between teachers and the governments and I would also recommend the provision of space and time for teachers to connect as peers. Echoing recommendations made by Rushton et al. (2017), I propose that both online and in-person communities focused on connecting physics teachers, particularly in areas with less dense populations, could offer this desired support. I contend this proposed community specifically designed for those teaching physics as opposed to science teachers in general. It may be small, but this community could promote connection sought by teachers in this study while providing a platform to share curricular innovations. This type of professional learning may better engage physics teachers as they can connect with a community asking similar questions and (Darling-Hammond & Richardson, 2009; Guskey, 2009), hopefully, lessening physics teachers' feelings of isolation.

I commend the Saskatchewan Ministry of Education for its commitment to involving teachers in the curriculum development process, but when it comes to implementing curriculum documents there is more work to be done. As many concerns shared in this study are not resolvable by teachers on a personal level, it now falls to those people implementing this curriculum and those people involved in the professional development of teachers to work with—and connect—teachers to address these concerns.

Acknowledgements I would like to acknowledge Drs. Gregory Thomas, Mijung Kim, and Lauren Sulz for their hours of support on refining this research. Further, I would like to acknowledge all my participants in providing their thoughts and time to this research.

References

Abadie, M., & Bista, K. (2018). Understanding the stages of concerns: Implementation of the common core state standards in Louisiana schools. *Journal of School Administration Research & Development, 3*(1), 57–66. https://doi.org/10.32674/jsard.v3i1.1932

Akalamkam, K. (2013). A framework for curriculum sequencing. *International Journal of Learning: Annual Review, 18*(12), 35–48. https://doi.org/10.18848/1447-9494/CGP/v18i12/47840

Anderson, S. (1997). Understanding teacher change: Revisiting the concerns based adoption model. *Curriculum Inquiry, 27*(3), 331–367. https://doi.org/10.1111/0362-6784.00057

Borgerding, L. A., Sadler, T. D., & Koroly, M. J. (2013). Teachers' concerns about biotechnology education. *Journal of Science Education & Technology, 22*(2), 133–147. https://doi.org/10.1007/s10956-012-9382-z

Braun, V., & Clarke, V. (2006). Using thematic analysis in psychology. *Qualitative Research in Psychology, 3*(2), 77–101. https://doi.org/10.1191/1478088706qp063oa

Braun, V., & Clarke, V. (2021). *Thematic analysis: A practical guide.* Sage Publications, Inc.

Council of Ministers of Education, Canada. (1997). *Common framework of science learning outcomes, K to 12: Pan-Canadian protocol for collaboration on school curriculum.* https://science.cmec.ca/framework/

Darling-Hammond, L., & Richardson, N. (2009). Teacher learning: What matters? [Research review]. *Educational Leadership, 66*(5), 46–53. https://www.ascd.org/el/articles/teacher-learning-what-matters

Fischer, C., McCoy, A., Foster, B., Eisenkraft, A., & Lawrenz, F. (2019). Use of the stages of concern questionnaire in a national top-down reform effort. *Teaching & Teacher Education, 80*, 13–26. https://doi.org/10.1016/j.tate.2018.12.019

Fuller, F. F. (1969). Concerns of teachers: A developmental conceptualization. *American Educational Research Journal, 6*(2), 207–226. https://doi.org/10.3102/00028312006002207

Gabby, S., Avargil, S., Herscovitz, O., & Dori, Y. J. (2017). The case of middle and high school chemistry teachers implementing technology: Using the concerns-based adoption model to assess change processes. *Chemistry Education Research & Practice, 18*(1), 214–232. https://doi.org/10.1039/C6RP00193A

Geng, J., Jong, M.S.-Y., & Chai, C. S. (2019). Hong Kong teachers' self-efficacy and concerns about STEM education [Special issue]. *Asia-Pacific Education Researcher, 28*(1), 35–45. https://doi.org/10.1007/s40299-018-0414-1

Goldstein, D. (2014). *The teacher wars: A history of America's most embattled profession.* Doubleday.

Guba, E. G., & Lincoln, Y. S. (1989). *Fourth generation evaluation.* Sage.

Gudyanga, R., & Jita, L. C. (2018). Mapping physical sciences teachers' concerns regarding the new curriculum in South Africa. *Issues in Educational Research, 28*(2), 405–421. http://www.iier.org.au/iier28/gudyanga.pdf

Guskey, T. R. (2009). Closing the knowledge gap on effective professional development. *Educational Horizons, 87*(4), 224–233. http://www.jstor.org/stable/42923773

Hall, G. E., & Hord, S. M. (2020). *Implementing change: Patterns, principles, and potholes* (5th ed.). Pearson.

Hall, G. E., Wallace, R. C., Jr., & Dossett, W. F. (1973). *A developmental conceptualization of the adoption process within education institutions* (ED095126). ERIC. http://files.eric.ed.gov/fulltext/ED095126.pdf

Ohlemann, S., Imhof, M., & Bellhäuser, H. (2023). Implementing reform in the teacher education system: Concerns of teacher educators. *Teaching and Teacher Education, 126*, 104087. https://doi.org/10.1016/j.tate.2023.104087

Kelly, A. M., & Sheppard, K. (2010). The relationship between the urban small schools movement and access to physics education. *Science Educator, 19*(1), 14–25. (EJ874150). ERIC. https://eric.ed.gov/?id=EJ874150

Kind, V., & Chan, K. K. H. (2017). Resolving the amalgam: Connecting pedagogical content knowledge, content knowledge and pedagogical knowledge. *International Journal of Science Education, 41*(7), 964–978. https://doi.org/10.1080/09500693.2019.1584931

Le Fevre, D. M. (2014). Barriers to implementing pedagogical change: The role of teachers' perceptions of risk. *Teaching & Teacher Education, 38*, 56–64. https://doi.org/10.1016/j.tate.2013.11.007

Limiansi, K., Suranto, P., & Setiawan, C. (2023). Biology teachers' perspective on change of curriculum policy: A case for implementation of "independent" curriculum. *The Qualitative Report, 28*(9), 2620–2638. https://doi.org/10.46743/2160-3715/2023.6204

Lowe, B., & Appleton, K. (2015). Surviving the implementation of a new science curriculum. *Research in Science Education, 45*(6), 841–866. https://doi.org/10.1007/s11165-014-9445-7

Mäntylä, T., & Nousiainen, M. (2014). Consolidating pre-service physics teachers' subject matter knowledge using didactical reconstructions. *Science & Education, 23*, 1583–1604. https://doi.org/10.1007/s11191-013-9657-7

Meltzer, D. E., & Otero, V. K. (2015). A brief history of physics education in the United States. *American Journal of Physics, 83*(5), 447–458. https://doi.org/10.1119/1.4902397

Molnar, T. A., Elliott, D., & McVittie, J. (2019). Science education in Saskatchewan: Inquiry and indigenizing. In C. D. Tippett & T. M. Milford (Eds.), *Science education in Canada: Consistencies, commonalities, and distinctions* (pp. 65–84). Springer. https://doi.org/10.1007/978-3-030-06191-3_4

Mulhall, P., & Gunstone, R. (2008). Views about physics held by physics teachers with differing approaches to teaching physics. *Research in Science Education, 38*(4), 435–462. https://doi.org/10.1007/s11165-007-9057-6

Nehmeh, G., & Kelly, A. M. (2018). Urban science teachers in isolation: Challenges, resilience, and adaptive action. *Journal of Science Teacher Education, 29*(6), 527–549. https://doi.org/10.1080/1046560X.2018.1474425

Ramnarain, U., & Fortus, D. (2013). South African physical sciences teachers' perceptions of new content in a revised curriculum. *South African Journal of Education, 33*(1), 15 p., Article 573. https://doi.org/10.15700/saje.v33n1a573

Robinson, S. A. (2006). Curriculum change in the 1980s: *Directions* and the core curriculum. In B. W. Noonan, D. M. Hallman, & M. P. Scharf (Eds.), *A history of education in Saskatchewan: Selected readings* (pp. 209–219). University of Regina Press.

Roehrig, G. (2023). Research on teacher professional development programs in science. In N. G. Lederman, D. L. Zeidler, & J. S. Lederman (Eds.), *Handbook of Research on Science Education (Volume III)* (pp. 1197–1220). Routledge.

Rushton, G. T., Rosengrant, D., Dewar, A., Shah, L., Ray, H. E., Sheppard, K., & Watanabe, L. (2017). Towards a high quality high school work force: A longitudinal, demographic analysis of U.S. public school physics teachers. *Physical Review Physics Education Research, 13*(2). https://doi.org/10.1103/PhysRevPhysEducRes.13.020122

Ryder, J., Banner, I., & Homer, M. (2014). Teachers' experiences of science curriculum reform. *School Science Review, 95*(352), 126–130. (EJ1032518). ERIC.

Saskatchewan Education. (1992). *Science: A curriculum guide for Physics 20/30.* Author.

Saskatchewan Ministry of Education. (2010). *Renewed curricula: Understanding outcomes.* https://resourcebank.ca/courses/renewed-curricula-understanding-outcomes

Saskatchewan Ministry of Education. (2016). *Physics 30 curriculum document.* https://curriculum.gov.sk.ca/CurriculumHome?id=72

Saskatchewan Teachers' Federation. (2023). *Governance handbook: Constitution and policies.* Accessed March 26, 2024 from https://www.stf.sk.ca/resource/stf-governance-handbook/

Saskatchewan Teachers' Federation. (2018). *Listen, learn, lead: Re-imagine education.* https://www.reimagineeducation.ca/

Saskatchewan Teachers' Federation. (2019). *Education re-imagined: 12 actions for education.* https://www.reimagineeducation.ca/

Sun, C., Raptis, H., & Weaver, A. (2015). Crowding the curriculum? Changes to grades 9 and 10 science in British Columbia, 1920–2014. *Canadian Journal of Education/Revue canadienne de l'éducation, 38*(3), 31 p. Article 1834. https://journals.sfu.ca/cje/index.php/cje-rce/article/view/1834

Sunal, D., Simon, M., Sunal, C., Ogodo, J., Harrell, J., & Aggarwal, M. (2019). Effects of professional development on reform in high school physics teaching. In D. Sunal, C. Sunal, & E. Wright (Eds.), *Physics teaching and learning: Challenging the paradigm* (pp. 21–56). Information Age Publishing.

Tesfaye, C. L., & White, S. (2012). Challenges high school teachers face. AIP Statistical Research Center. https://www.aip.org/sites/default/files/statistics/highschool/hsteacherchall-09.pdf

Tytler, R. (2010). Stories of reform in science education: Commentary on opp(reg)ressive policies and tempered radicals. *Cultural Studies of Science Education, 5*(4), 967–976. https://doi.org/10.1007/s11422-010-9277-3

van den Berg, R., Sleegers, P., Geijsel, F., & Vandenberghe, R. (2000). Implementation of an innovation: Meeting the concerns of teachers. *Studies in Educational Evaluation, 26*(4), 331–350. https://doi.org/10.1016/S0191-491X(00)00022-5

Wallace, C. S., & Priestley, M. R. (2017). Secondary science teachers as curriculum makers: Mapping and designing Scotland's new curriculum for excellence. *Journal of Research in Science Teaching, 54*(3), 324–349. https://doi.org/10.1002/tea.21346

Watson, E., & Rose, K. (2019, June 3). *How many participants can I have? A collaborative autoethnography examining graduate student access to teacher participants* [Paper presentation]. Canadian Society for the Study of Education Annual Conference, Vancouver, BC, Canada.

Wei, B., & Chen, N. (2019). Agency at work: Two beginning science teachers' stories in a context of reform in China. *International Journal of Science Education, 41*(10), 1287–1302. https://doi.org/10.1080/09500693.2019.1600205

Wheelahan, L. (2010). Why knowledge matters in curriculum: A social realist argument. *Routledge.* https://doi.org/10.4324/9780203860236

Lau, W. W. F., & Jong, M. S. Y. (2023). Typology of teachers' stages of concern for STEM education. *Research in Science & Technological Education, 41*(4), 1560–1578. https://doi.org/10.1080/02635143.2022.2064447

Yan, T., & Deng, M. (2019). Regular education teachers' concerns on inclusive education in China from the perspective of concerns-based adoption model. *International Journal of Inclusive Education, 23*(4), 384–404. https://doi.org/10.1080/13603116.2018.1435741

Chapter 3
The Vital Links Among Science Pedagogical Awareness, Emergency Preparedness, and Creativity in Crisis: When the Rulebook is Constantly Under Revision

Katarin A. MacLeod and Wendy L. Kraglund-Gauthier

3.1 Introduction

The abrupt and ongoing cyclic closures of postsecondary campuses and, in fact, entire educational systems in response to the SARS-CoV-2 pandemic left the traditional teaching and learning system of course and program iterations in upheaval. At the beginning of the pandemic, university instructors were thrown into planning and preparing for emergency remote teaching (ERT), "a temporary shift of instructional delivery to an alternate delivery mode due to crisis circumstances" (Hodges et al., 2020, p. 6). At StFX, Katarin, along with many of her colleagues in Nova Scotia and across Canada, was tasked with moving all courses and programs that until then were designed and delivered face-to-face to an online format. The transition to ERT was swift across the country, and most instructors experienced a steep learning curve associated with online platforms and changes to pedagogy. Katarin's courses involved hands-on laboratory experiments and activities with preservice science teachers, which made planning and implementing lessons and assessing learning more difficult as well as safely planning and performing experiments while online. Wendy had less of a transition because she was teaching in a fully online MEd program and, as an instructional designer, was accustomed to exploring the pedagogical implications of online course delivery. However, her students were dealing with their own transition

K. A. MacLeod (✉)
St. Francis Xavier University, Antigonish, NS, Canada
e-mail: kamacleod@stfx.ca

W. L. Kraglund-Gauthier
Yorkville University, Fredericton, NB, Canada
e-mail: wkraglund-gauthier@yorkvilleu.ca

C. D. Tippett and T. M. Milford (eds.), *Seeing Science Through the Eyes of Canadian Teachers and Learners*, Contemporary Trends and Issues in Science Education 54,
https://doi.org/10.1007/978-3-031-88785-7_3

to ERT, which manifested in their heightened personal stress and Wendy questioning how she approached online learning.

Depending on the instructional mode that they are required to use, university instructors have varying levels of experience and comfort with designing and delivering engaging learning experiences. Courses may be fully online as synchronous or asynchronous or blended, that is, with some components happening in a physical classroom and others online. Courses may be hybrid with some students attending in a physical space while others attend online at the same time. Finally, courses may be entirely face-to-face with all students physically present and no online option available. When the pandemic began, some instructors were already proficient in an online environment. Some instructors improved their online teaching practices with access to pedagogical support and mentorship from instructional designers, campus teaching and learning centres, or through mentoring by colleagues. However, other instructors struggled to find adequate and appropriate supports.

These teaching and support scenarios existed before, during, and after the pandemic. Although pandemic restrictions eventually relaxed and classrooms filled again, educational systems may face another pandemic, an environmental disaster, or an insurrection that will force another rapid adjustment to teaching. The better prepared instructors are for such scenarios, the less disruptive these adjustments will be for all involved. For instance, schoolteachers have their students practice fire drills so that they know what to do, can help each other, and get through an emergency in a safe, calm, and organized manner. We can keep this analogy in mind as we approach ERT at the postsecondary level as pedagogical preparedness: planned and organized approaches to instruction so that educational standards are upheld and learning outcomes achieved.

Key to understanding the pedagogical implications of ERT is its emphasis on temporary solutions, which suggests less planning and preparation without fulsome attention to the diverse learning needs of students and instructors. For example, after surveying 208 postsecondary chemistry students at the onset of the pandemic, Jeffery and Bauer (2020) noted that "the stress of events revealed and amplified individual differences and the importance, for any educational setting (in-person or online), of cognitive processing limitations, social dynamics, peer interaction, real-time discourse and visualization of problem-solving processes, and hands-on manipulation" (p. 2472). Instructors need to be aware of these multifaceted challenges and learn how to meet them.

In this chapter, we share our ERT experiences, which may differ from the experiences of colleagues at other institutions due to our situatedness during the pandemic, juxtaposed with our specific issues of assessment, accessibility, instructional design, and content. This self-study is storied through our experiences as educators and researchers of preservice teacher education. Katarin has spent more than 20 years as a science/physics/STEM educator in the secondary school system and as a professor of undergraduate physics and in teacher education with a focus on science literacy and assessment. She has taught both online and in a variety of blended modalities with experience in both synchronous and asynchronous delivery of course content. Wendy has spent more than 25 years in postsecondary contexts as a face-to-face, online, and hybrid educator and as an instructional designer focused on accessible online pedagogy. When the pandemic was declared, Wendy assisted Katarin in the redesign of her secondary science methods course as the instructional model shifted with institutional pandemic restrictions. Instruction varied between fully online, fully face-to-face with COVID-19 restrictions, hybrid, and blended—each learning environment required pedagogical workarounds for a multitude of situations.

Moving to an online teaching environment requires changes—not necessarily in terms of philosophical underpinnings of teaching and learning but rather in the ways that one's pedagogical stance shifts in virtual spaces (Kraglund-Gauthier, 2020). For some instructors, this shift necessitates a minor change; for other instructors, the shift is more substantial. Here, we have laid bare our revelations about teaching practices and access to learning that the pandemic brought to the fore, identifying the benefits, challenges, and unexpected outcomes.

3.2 Literature Review

We share pedagogical insights that emerged while we navigated an unprecedented time in history, with the intent of informing future practice for ourselves and others. Our insights are made in the context of teaching and learning online with a focus on the supports needed to facilitate technological, pedagogical, and content knowledge (TPACK; Koehler & Mishra, 2009). It is the fusion of these two areas along with our lived experiences that makes space for these stories to unfold.

3.2.1 Teaching and Learning Online

There is research on the effectiveness of teaching and learning online with respect to achieving learning outcomes, yet the degree of effectiveness is dependent on various factors, including the nature of the course itself, the quality of the learning design, and students' motivation to learn and their preparedness to learn in a digital space (Guevara et al., 2021; Hu et al., 2021; Kraglund-Gauthier, 2020). Additionally, online delivery can reduce some barriers to learning while amplifying others. Online delivery enables access to courses and resources from anywhere in the world, but this advantage is tempered by (a) equitable access to Internet bandwidth and devices and (b) adherence to accessibility standards (Bates, 2022; Jeffery & Bauer, 2020). Bates (2022) referred to the ongoing debate over the superiority of one instructional mode over another, noting with an assumption that face-to-face is the most effective mode. However, as others (e.g., Kraglund-Gauthier, 2020; Kwon et al., 2019; Pandey & Mishra, 2021) have explained, the determination of effectiveness is rooted in both the educator's and their students' level of digital literacy and the educators' understanding of effective pedagogical practices in online contexts.

Following on their collaborations with scholars in the fields of higher education policy, administration, and economics, Chickering and Gamson (1987) listed seven principles for good pedagogical practices:

1. Encourages contacts between students and faculty.
2. Develops reciprocity and cooperation among students.
3. Uses active learning techniques.
4. Gives prompt feedback.
5. Emphasizes time on task.
6. Communicates high expectations.
7. Respects diverse talents and ways of learning. (p. 2)

Although these principles were published more than 30 years ago and developed from research on face-to-face classroom, they are still useful for the online learning environment. For example, instructors who focus on constructivist pedagogical activities can efficiently maximize students' engagement and motivation; in turn, students may feel a sense of connection with instructors, classmates, and content—regardless of the mode of teaching and learning (Boettcher & Conrad, 2021; Kraglund-Gauthier, 2020; Miller, 2020). However, the online mode of teaching and learning requires different approaches to content, alternate mechanisms for student engagement, and other forms of assessment (Bates, 2022; Dziuban et al., 2018; Efwinda & Mannan, 2021; Esfijani & Zamani, 2020).

Our focus for this chapter is the blended learning mode where we use Bates' (2022) broad definition of a portion of face-to-face instruction supplemented with asynchronous components that increase the flexibility of course delivery and learning tools while maintaining interactions between learners and instructor. A flexible course design can help to meet the needs of adult learners, (Dziuban et al., 2018), especially needs outside the academic sphere, for example, scheduling, health, or transportation issues (Kraglund-Gauthier, 2020). With a blended mode, some students may participate more online while others may participate more during face-to-face classes, demonstrating how flexibility in the mode can benefit students (Lai et al., 2016).

The teacher education literature reveals a shift from lecture to student-centred instruction in which students become active and interactive learners throughout the entire course, including in-person classes and other delivery methods such as blended learning (Andrade & Coutinho, 2017; Burns et al., 2020; Dziuban et al., 2018). Over the past 20 years, the higher education classroom experience has also undergone a shift; specifically, blended learning has resulted in transformed practices at many institutions of higher education (Dziuban et al., 2018). In the self-study described in this chapter, the pedagogical approaches utilized in the blended mode combined the social opportunities of the classroom (Jeffery & Bauer, 2020) with the learning possibilities afforded by the online environment (Dziuban et al., 2018; Lai et al., 2016). One model that can support teaching and learning in the blended mode is TPACK (Koehler & Mishra, 2009).

3.2.2 Support for TPACK

Koehler and Mishra's (2009) TPACK model is an extension of Shulman's (1986) work describing an educator's evolution from content knowledge—CK and pedagogical knowledge—PK to pedagogical content knowledge—PCK. Taken as a whole, the TPACK model serves to underscore the challenges of teaching with technology and reveals "the types of flexible knowledge needed to successfully integrate technology use into teaching" (Koehler & Mishra, 2009, p. 60). Each TPACK element can be considered separately, yet effective integration of technology in learning design involves the purposeful weaving of all three elements together while taking the needs of the student into consideration (MacLeod et al., 2022).

Online and blended course design needs to be carefully considered, and instructors must learn to use the technical tools available and to navigate a course learning management system. The development of training for technology integration can be daunting as technology is constantly changing (Akcil et al., 2021), and research on the effect of training has been mixed. Some researchers have found that professional development and teaching and learning supports do not have a significant influence on (a) instructor self-efficacy or beliefs around technology and (b) the quality or quantity of information and communications technology integration (Barton & Dexter, 2020; Esfijani & Zamani, 2020; Kwon et al., 2019). In contrast, other studies have revealed higher self-perceptions and increases in self-efficacy along with skill acquisition after attending training (Alhassan, 2017; Canbay, 2020; Efwinda & Mannan, 2021; Kraglund-Gauthier & Moseley, 2019).

However, support for pedagogical and technical skills is important and, as such, is available at many postsecondary institutions through teaching and learning centres. Yet these centres are often only open during business hours, which limits accessibility, particularly for adjunct instructors with alternate day jobs and instructors teaching evening courses (Kraglund-Gauthier & Moseley, 2019). Additionally, accessing these supports is usually optional; when making decisions on how to spend their time, instructors may not prioritize their own learning needs (Kraglund-Gauthier, 2020) despite rapidly changing teaching environments and differences in existing technology (Akcil et al., 2021; Rodríguez-Muñiz et al., 2021).

3.2.3 Research Context

When Canadian universities began their winter terms in January 2020, it was business as usual. At Katarin's university in Nova Scotia, students arrived on campus and began their mostly face-to-face classes and laboratories. On January 4, 2020, the World Health Organization reported a cluster of pneumonia cases in Wuhan, China; by January 30, 7,818 cases of a novel corona virus outbreak were reported worldwide (World Health Organization, 2020). By February 11, this novel virus had a name: COVID-19.

On March 11, the World Health Organization declared a pandemic and life changed abruptly. International and domestic borders were closed, and public transportation halted. At Katarin's institution, students and instructors were told to go home for a few weeks until the emergency was over. On March 22, the Government of Nova Scotia (2020) declared a state of emergency that immediately shuttered businesses, restricted movement, and closed schools. Noting the significance of this change, Katarin and Wendy embarked on a qualitative case study of their experiences.

3.3 Method

This qualitative case study was focused on developing understanding our experiences and contributing to the "knowledge base and practice of education" (Merriam, 1998, p. 1). A case study is a way to capture the "circumstances and conditions of an everyday situation" (Yin, 2018, p. 51). The specific situation here was our teaching of preservice and inservice teachers during the first year of the pandemic in which we reflected on our pedagogy as it was affected by health and safety restrictions. The data are narrative reflections made during the academic year that we subsequently examined to explore intersections and divergences. From a subjective epistemological position (Leavy, 2023), we conducted a self-study, making ourselves "the case under investigation (Leavy, 2023, p. 163) because of its potential to help us align our teaching intentions with our teaching practices. Our use of vignettes (i.e., narratives or storied accounts) as data was motivated by our focus on praxis and desire to explicate our subjective and transformational experiences for other practitioner-researchers (Skilling & Stylianides, 2020).

Our vignettes of "specific situations and the problems that might arise in them" (Torres, 2009, p. 94) were written independently based on our reflections and class journals capturing pivotal moments from the 2020–2021 academic year. The vignettes depict real-life events and reflections that have then been turned into a story (Skilling & Stylianides, 2020) and are used to explore the overall benefits, challenges, and unexpected outcomes that emerged during ERT. They offer glimpses of the links between choice—or lack thereof—of teaching and learning mode and self-efficacy (Bandura, 1993). They also capture the impacts of teaching in our required versus preferred mode on our motivation, and our perceptions of student and administrator attitudes toward our navigation of ERT.

3.3.1 Analysis and Discussion

Taking a constant comparative approach (Glaser, 1965), for the first round each author independently read the other's vignettes line by line and identified codes or key terms that connected to the broad categories of benefits, challenges, and unexpected outcomes. These open codes and key themes were tracked across vignettes, and memos were written to gain insight. Then we compared the codes and key themes and discussed the memos and meanings that we had generated from the vignettes. This approach provided clarification and confirmed that we had reached saturation in our analysis of the vignettes. From this analysis, themes emerged that represented noteworthy and recurring findings.

In the following section, we present a discussion of the benefits, challenges, and unexpected outcomes in moving to remote teaching during a global pandemic. This discussion is presented holistically through the story-telling vignettes written during the 2020–2021 academic year, which reveal the authors' emergent understandings and connections to theory and praxis.

3.3.1.1 When the World Shut Down

When the life-threatening effects of the COVID-19 virus became evident, governments around the world made the unprecedented decision to close their borders and their physical spaces. In Canada, by acts enforced via various emergency measures at the provincial and federal levels, only individuals working in essential services were permitted to interact with others, albeit with strict safety protocols. Postsecondary institutions were abruptly thrust into unknown territory where their face-to-face systems and programs quickly pivoted to a virtual space. The challenges of this pivot were great; instructors struggled to adjust their pedagogy to function in often unfamiliar, digital, asynchronous, and synchronous spaces. Moreover, as Burns et al. (2020) observed, "developing an online practicum experience for preservice teachers was not even on the radar of most teacher education programs before March 2020" (p. 6).

In the first vignette, Katarin reflects on her secondary science methods class from the winter 2020 semester that was designed as an on-campus, face-to-face course with a laboratory component. She includes reflections on the transition the university made when pandemic restrictions were imposed and on what the practicum component (i.e., practice teaching for student teachers) was to look like given the new restrictions.

> *Katarin:* During the winter semester of 2020, as COVID-19 spread, my science education students and I talked about the virus, its spread, what we were learning about the level of scientific literacy amongst the general population, and what their role was as science educators. We talked about what their practicum placements might look like in this new and changing context. The term's course-based components had just ended on March 6, and my teaching assignment with them had concluded. I was marking assignments, and students were preparing for their practicum placements that were scheduled to begin after the provincial school system's annual break on March 9–13. Then, the university announced that it was closing due to COVID-19 and the education students' March break would become 2 weeks instead of 1 week. While the students were on break, faculty members worked to salvage what should have been a face-to-face practicum session but could not be. Travel and health restrictions within the province meant preservice teachers were not allowed into the schools.

In the next vignette, Wendy reflects on her position in pre-pandemic winter 2020 as an on-campus, part-time instructional designer working with faculty members to transition their courses from face-to-face to online.

> *Wendy:* Ironically, in late February 2020, I left the university where I worked part time as an instructional designer responsible for helping faculty members transition their courses online. Admittedly, I had become a bit frustrated with the slow pace of adoption and institutional commitment to online learning. Over the years, I had worked with numerous faculty members and community-based organizations as they transitioned from face-to-face to online environments to meet and deliver content; I was confident in how web-based tools could be used to connect, collaborate, and learn. I have led and participated in research and offered professional development in how learning outcomes can be achieved in a digital space, and I have advocated for equity and accessibility regardless of mode. I am a realist, however, who recognizes that shifting to online delivery is not an easy transition for some individuals for a myriad of reasons. A digital classroom is not the same as a physical classroom and should not be considered as such.

In both vignettes, it is clear that the authors were aware of the role online teaching and learning can play in the education system, be it at the K–12 level or in higher education. Further, it is clear that both were aware that change of teaching mode takes considerable planning. It was also clear that something significant was about to happen that was well beyond their control.

Katarin explains her experiences when emergency measures were enforced: all non-essential work ceased, schools closed, and student teacher practicums were cancelled by the provincial government. Still bound by teacher certification requirements, the university's Faculty of Education was thrust into the work of designing an online practicum assignment for the student teachers to complete what would be a reasonable replacement for in-class experiences.

> *Katarin:* As a Faculty of Education, we had to pivot and create an alternate online practicum experience that mapped practicum learning intentions with assignments over three phases that preservice teachers could complete while working from home. In Phase 1, preservice teachers were to develop a 2-week, in-depth unit of study that could be delivered in a face-to-face classroom. Then, in Phase 2, they were to modify their unit for online learning or an at-home learning/correspondence unit. Finally, in Phase 3, they were to write about the challenges, benefits, and learnings of the activity. All this was essentially independent work—a stark departure from the supportive practicum programming typical of prepandemic teaching placements and supervision.
>
> We were adrift, with no navigational charts to guide our journey. It was a difficult time for all—dealing with the uncertainty in the world, concerns over their health and the health of their loved ones, and the practicalities associated with completing the mandatory requirements for degree eligibility and teacher licensing. Supporting preservice teachers took on an entirely new meaning that extended beyond supporting learning outcomes to include navigating the digital space to meet in real time and to connect and share documents. These necessary things—meeting online, sharing documents, helping work through the assignment—needed to be done on a one-on-one basis due to schedules, connectivity and device limitations, employment, and healthcare schedules.

This alternate practicum assignment appeared straightforward and pedagogically sound to Katarin in her role as a faculty advisor. Yet supporting her preservice teachers in meeting the learning outcomes of this assignment took a substantial amount of time and mentoring because many of the preservice teachers had little practical experience in how to use, plan for, or connect content with available various digital platforms. Only a few preservice teachers had ever considered the infusion of technology into their lessons, let alone a fully online lesson. The learning curve was steep, and it was just the first COVID-19 wave.

Both Katarin and Wendy knew the preservice teachers' transition to ERT and online teaching was going to be difficult based on the varied levels of TPACK awareness exhibited in face-to-face BEd classes prior to practicum. Based on student reports, many were experiencing their first online course—a teaching practicum assignment where they had to create online lessons with no previous experience or resources. Additionally, they faced barriers related to inadequate access to appropriate devices, the cost of Internet data, and even reliable Internet connections for those who lived in areas with limited or unreliable service. Wendy and Katarin discussed these issues at length, both anticipating and troubleshooting in the moment, creating individualized plans for each of Katarin's practicum candidates according to their situational needs.

Next, Wendy reflects on her experiences teaching within the Faculty of Education at Yorkville University, via online delivery in the winter 2020 semester and during the shutdown event.

> *Wendy:* Those first few months of lockdown were a blur. At Yorkville, our MEd program has always been delivered online, and instructors worked at a distance, so we had no need to adjust our pedagogical delivery. What shifted, however, was the level of anxiety in our graduate students, who found themselves shifting from being online students learning how to teach in face-to-face classrooms and workplace learning contexts to designing their own online programs for ERT, often competing with their children for device access and Internet band width. My inboxes became filled with email messages that typically started with "I imagine you're really busy with other folks right now, but I really need your help to …" ERT meant a scramble for quick fixes in the short term; but as the pandemic dragged on, there was a clear need to think more deeply about content and course design, about assessment and accessibility—especially in hands-on learning contexts such as science methods courses. Understanding the intersections of TPACK on teaching and learning became paramount.

These two vignettes focus on the postsecondary students and their ability to navigate their course work during the first few months of the pandemic. Although they knew how to function in their regular classroom, whether that be a face-to-face practicum or an online education class, the additional stressors and complications associated with the pandemic made the *usual* seem *nearly impossible*. Despite the challenges from this first COVID wave and recognizing that the implementation of ERT measures was a quick fix, those fixes did get these students through the winter 2020 semester and a required teaching practicum as well as a virtual convocation and provincial licensing for the graduating students. Yet Katarin and Wendy were left asking themselves deeper, critical questions about not only ERT but also the preparedness of teachers to function in a digital world.

As September neared, Katarin had to change gears again. Her institution decided to bring most students back to campus and into a face-to-face learning environment. The return-to-campus decision was not made lightly. Stakeholders had to commit to approaches that, upon implementation, would keep everyone as safe as possible while ensuring content delivery standards were met and learning outcomes achieved.

Katarin: During a pandemic in a face-to-face [F2F] learning environment, exactly how does one prepare to teach a curriculum and instruction course for secondary science preservice teachers who do hands-on learning when all your colleagues are telling you about their online learning strategies? Who do you turn to since there is no one else in "your boat?" When I tried to explain my situation to others, the response was "Oh, you are so lucky!" or "Really?" This version of F2F was not exactly F2F but a quasi-blended F2F sort-of model ... sometimes.

Wendy: Okay, so what is required and what is extra? As you figured out, everything will take longer online ... now it's back to face-to-face, but the setup, assessment, and teardown is more complicated. How about instead of learning stations you move to learning pods, where the activities and not the people rotate? What can be "flipped" and done outside of class? What's the Plan B if the class becomes an exposure site?

As the vignettes reveal, both authors had extensive foundational pedagogical education in face-to-face contexts and experience in online course design and delivery. Yet, like her preservice teachers, Katarin's transition to an online format was not easy or smooth. The issues she faced went beyond the content of the curriculum or the mode of delivery; rather, they echoed those discussed by Buksh (2020), who argued that students first needed their individual and personal needs supported before they could approach their learning and pedagogical development.

3.3.1.2 Transitioning to Online and Modeling Pedagogy

Instructor, student, content, and setting each play a key role in the development of learning environments in which the skill of instructors is often measured in terms of their ability to create activities that attend to their students' specific learning needs. For this reason, pedagogy is contextually bound to learners and their educative setting (Leach & Moon, 2009). Within a teacher education program, the concept of pedagogy takes on import because modeling effective teaching practices entails critical thinking, reflection, self-study, and a focused analysis on the issues that shape pedagogical practice (Loughran, 2007).

Boettcher and Conrad (2021) stated that the most important element when transitioning to online learning in higher education is ensuring the educator has a proper foundation in principles and practices in their own teaching. From a pedagogical standpoint, including Chickering and Gamson's (1987) foundational principles of effective teaching into the design, delivery, and assessment of learning outcomes in an online classroom is a prudent decision. When thrust into ERT, this need became abundantly clear.

Katarin: There is an adage that "good" teaching looks effortless. In reality, it is hard work, a process of learning from trial-and-error, applying theory, and reflexive thinking. In a preservice teaching program, pedagogy is made explicit. In my own classrooms, we run through science lessons, but then break them down in terms of pedagogical processes and outcomes. Students benefit from their time engaging with me, their peers, and the content, trying to find their own pedagogical process that "feels" right for them. Practicum is a time to test it out, reflect on their learning, and get feedback from their supervisors.

Wendy: Thinking back to some of the conversations I have had with faculty [members], not only now in this disruptive pandemic space but also in planned transitions to online delivery, coming to instructional design with a teaching background has certainly been a benefit. Before we can get to the "how to" of a technological tool, I asked faculty [members] to consider the "what and the why" of their thinking. Too often, technology drives decisions for learning design. This is backwards. Learning design needs to inform decisions on what tools are appropriate to achieve learning outcomes.

Moore (1989) identified three types of student learning interactions in distance education: student–student, student–content, and student–instructor that have since formed the basis of pedagogical engagement online (Abrami et al., 2011; Pandey & Mishra, 2021). How instructors engage their students is due, in part, to the creation of spaces conducive to exploration and experimentation, which draws on prior knowledge and experience and an opportunity to share that knowledge and experience and learn from others. With advances in technology since Moore's writing, the possibilities of how these interactions happen in online spaces requires a certain mind-shift away from instructor-led didactic methods toward more collaborative activities and reflective applications of knowing and understanding. In an online environment, these elements take on additional import because of the transactional distance between individuals (Friesen, 2011) and advances in the very technology that connect individuals in a virtual space.

In the design process, the instructor's pedagogical decisions involve identifying content sequence, creating activities that demonstrate student learning, and facilitating how students acquire content knowledge (Hunter, 2004). When an instructor designs an online course, regardless of being in an emergency or not, the task is not to change the on-campus course to suit the online learners; rather, it is to create a new distance learning environment. It involves creating a community of learners who may never meet face-to-face. It also requires knowing a bit more about the campus infrastructure; for example, how to access library resources at a distance, how assessment *for* learning and assessment *of* learning can be done at a distance, and how to create and monitor collaborative group work. It entails creating a community of practice (Wenger, 2006) online. Moreover, as the Columbia University Center for Teaching and Learning (2020) argued, instructors need to identify key elements that increase positive student learning throughout their transition to blended learning spaces. This process requires acknowledgement that students may need explicit guidance in how to appropriately use digital tools for engagement (Abrami et al., 2011; Burns et al., 2020) and that their learning curve may be just as steep as that of their instructor.

In Katarin's context, this explicit guidance was both a benefit and a challenge in that it afforded preservice teachers the opportunity to learn and practice skills that they would not normally have acquired. However, the time it took to integrate the technology into the classroom learning at the beginning of the semester was an eye-opening experience for all involved due to the limited technical experience of the preservice teachers. Ultimately, the results were positive, because all preservice teachers achieved additional skills in navigating the technology and then utilized those skills during their practicum.

3.3.1.3 Drawing from Past Learnings

It has been well documented that an individual's past experiences, prior knowledge, and personal views and opinions tend to impact the type of activities required for learning (Barton & Dexter, 2020; Birisci & Kul, 2019; Crawford, 1996; Efwinda & Mannan, 2021; Vygotsky, 1978) and that "their interpretations of the purposes or goals of an activity" (Crawford, 1996, p. 44) are critical for the learning experience. Additionally, the investment of time needed to learn new techniques for content delivery and ways to engage students and ever-changing technical elements of software can create frustration because it is time taken from preparing content and being responsive to students' needs (Allen & Seaman, 2012; Kraglund-Gauthier & Moseley, 2019). The question of how to integrate technology into the pedagogical process to manifest students' deeper, richer cognitive understanding and how to procedurally facilitate this, given limited resources and time, is a struggle for many educators.

As educators ourselves, we recognize that familiarity and early use are the most challenging first steps in the process of technological integration (Rogers, 2011) and the point at which a great deal of frustration and decreased self-efficacy can occur. Despite best intentions, unfamiliarity with the tools and lack of confidence can lead instructors to default to a technology-enhanced lecture method (Rogers, 2011). Often, one gravitates toward that which one knows and feels comfortable, falling back to teaching how one was taught. The following vignettes reveal the challenges in finding one's pedagogical stance and the benefits of having supports in place.

> *Katarin:* The time that it takes to learn and understand is time well spent, or so the saying goes; and this is the case for integrating technology into classroom practice during a pandemic or during "normal" times. Understanding the tools that you have to help you teach and provide programming is critical, just because some of those tools might be in a black box and you don't exactly understand the details of how they work doesn't mean that you can't make them work for you and your students. This was where we started when we began looking at what was "under the hood" and available to us for software in the fall of 2020 to help us connect, communicate, and teach. What do we already have, how do we use it, how can we use it better, and let's learn here so that when we go out on practicum we know what we are doing. This sense of "we are all in this together and we can learn from each other" seemed to work very well. When we had questions, we asked each other; someone had tried it; someone usually had a work around or had contacted IT [information technology support], so we had a solution. Thirty-four people working toward the same end goal makes the magic happen faster than just one person.

> *Wendy:* In my doctoral research, I explored the changes face-to-face educators experienced when they transitioned to online teaching. Their academic identity was wrapped up into how they functioned in a physical classroom space and interacted with students. They worried that, if students saw them fumbling with technology, they would not be seen as the content expert either. This was a familiar refrain also heard in my instructional design work both prior to and during the pandemic. With so many "bells and whistles" to choose from, they either

> stuck with the safe and familiar or tried to do too much—overly complicating the learning experience. And then, when it went sideways, I had to step in and help them determine why and how to fix it for the next time, rather than give up.

If educators see technology modeled effectively as an everyday pedagogical practice in learning spaces, they are more apt to take on new technological challenges themselves (Kraglund-Gauthier, 2020; MacLeod & Kraglund-Gauthier, 2015). The Columbia University Center for Teaching and Learning (2020) reported that most educators' mistakes involved trying to move their programs online without taking the time to rethink assessments, learning experiences, and course goals. Yet when an attitude of emergency preparedness is embedded into planning and delivering lessons, the groundwork and familiarity of the alternative method can be built into the existing course; educators and students are ready for ERT, and it is less of a crisis scenario. This was our goal for the science course that students be ready for ERT because of the groundwork.

Functioning effectively in digital spaces requires skillful integration and modelling of each identified competency of teacher education. It requires recognition that the appropriate application of technology is pedagogically linked to the course content. "Transitioning to an online environment requires the reinvention, reimagining, and applying of 'e-flavors' of general learning theory" (Guevara et al., 2021, p. 1). Specific competencies in creating and demonstrating techniques and communicating effectively during class time are required so that problems can be minimized. Integrating technology is not without inherent risks; considering that "problems are often seen as an indicator of incompetence and failure" (Osterman & Kottkamp, 1993, p. 21), it is not surprising that teacher educators—who have the role and responsibility to model effective practice—are reluctant to try. Further, as Boettcher and Conrad (2021) stated, the most important element when transitioning to online learning in higher education is ensuring the educator has a solid grounding in pedagogical understanding. Answering the question of what digital tools to use to enhance the science lesson is critical rather than the tool being an add-on.

In an ERT context, despite the need to expedite decision-making (Burns et al., 2020), pedagogical awareness needs to inform preparation. Emergency preparedness is a term that, depending on the audience, can have many different levels of meaning. In Canada, we operate under Public Safety Canada (2022), which:

> works in collaboration with other federal departments and provincial and territorial governments to strengthen national emergency preparedness, [where] these efforts, taken prior to an emergency, help support event response, contribute to reductions in the impacts of events and help identify opportunities for future prevention and mitigation efforts. (Emergency Preparedness section)

Moreover, the concept of resilience is defined as "the capacity of a system, community or society to adapt to disturbances resulting from hazards by persevering, recuperating or changing to reach and maintain an acceptable level of functioning" (Public Safety Canada, 2019, p. 5). This capacity is exactly what is needed in ERT, as the authors' reflections reveal.

Katarin: In early September 2020, we were allowed back on campus. On my first day, as I stood at the back of my empty lecture hall, I wept. Forty individual tables with chairs arranged six feet apart in a perfect set of rows and columns. It took me back to the mid-1990s. The new rules meant there was to be no movement of desks and no close discussion among students. I felt as if my essence as a pedagogue had been destroyed. I am *that professor* who encourages students to ask questions, discuss, work in groups to solve problems; we make noise and interact with each other. Would students follow the protocols? Would I be loud enough? How would this ever work? Well, I needed to make it work. That's what physicists do: we solve problems, and this was a good one, albeit a logistical nightmare.

So, I thought back to how I encouraged my preservice students to invest time demonstrating, repeating, being understanding, and explaining so that everyone always got to the end goal. Starting on Day 1, I needed to show them explicitly how the classroom rules and etiquette were different from what they were accustomed. Therefore, I greeted each of them, reminded each to use hand sanitizer, to adjust their mask if necessary, and to sanitize their *island* (how we referred to their individual desk) rather than assuming the custodian had been able to do it between class changes. This new routine became the norm, no questions, no fuss, just friendly reminders from peers or me, and we moved on. With the routine, students seemed to begin to relax as they entered the room; they knew the expectations and we had a classroom community and a COVID-19 protocol for our classroom that was enforced consistently. It was also a teachable moment—an opportunity to connect to their content knowledge of biology, spread of disease, and their emerging role as "the science teacher."

Wendy: Safety protocols in a science laboratory need to be second nature for all, and teachers need to remain calm when incidents happen. When we routinize certain aspects of our work, expectations and standards can remain high. Yet, it's hard to be the Protocol Police, whether that be enforcing pandemic health protocols or ensuring that the classroom—digital or otherwise—is accessible for learning and experimenting. It can be a life-or-death scenario or simply an effort to have all students feel safe in the learning space. It's human nature to seek shortcuts and to become lax when a certain comfort level is reached. Being mindful of the *why* is key. Enacting teaching protocols for safety—not just for chemicals but for emotions, too—and ensuring content is accessible to all learners is crucial. Can the student see the demonstration from their seat? Can they hear conversations muffled behind masks? Did they have the bandwidth, both technological and emotional, and time to review the lesson before class? Are they ready to learn?

Emergency preparedness is predicated on the notion of repetitive behaviour, practice, and remaining calm in crisis. Through repetitive behaviour, the new norm for this class led to a sense of calmness amidst the chaos in the world. Knowing they were entering a safe space to learn helped fulfill Maslow's foundational tenet of the need for safety; without safety, climbing Bloom's hierarchy from knowledge up to creative expression can be a struggle (Buksh, 2020). The challenge, however, was that the institutional academic leadership team set very broad parameters; therefore, it was up to individual professors to create the caring classroom learning space that could support the achievement of learning outcomes.

3.3.1.4 Generation Z: Meet Your Technology

In recent years, technology seems to have become all-pervasive; it is being used by most people in some shape or form to work, play, and learn. People have become digital creators and curators of content, benefiting from the ability to access online resources from practically anywhere at any time. Yet as the pandemic revealed, there are inequities in access that are exacerbated along demographic factors of race, ethnicity, gender, socioeconomics, and ability/disability. It can be challenging for educators to ensure that the tools they use, and expect students to use, are accessible to all.

Shortly after the fall 2020 semester began, Katarin used tools she had learned from Wendy years before while teaching graduate students. This time, students in Katarin's face-to-face science education course were introduced to basic and advanced features of MS Office 365 and Moodle (i.e., a free open-source, online learning system). Specifically, they were introduced to OneDrive for resource sharing; OneNote, where a class notebook was created where they could post their reflections and assignments; Outlook, for emails and calendars for meeting invitations and links; and advanced features of Word, PowerPoint, and Sway for document creation and presentations. Although they had used Moodle as a course document repository, they needed to learn the interactive features of the system from a teaching perspective in preparation for a practicum in which K–12 students were also learning via Moodle.

They also learned to use Blackboard Collaborate, the learning platform used to deliver synchronous online courses in the MEd program. As well as learning to be participants, they learned how to host so they were able navigate the space as a pedagogical tool for instruction. Using Blackboard Collaborate together in the face-to-face class meant that, with breakout rooms and individual headphones, students could have small-group conversations without leaving their island. They were able to engage in active science learning, that is, completing experiments, sharing observations, and discussing in real time while following protocols. Document sharing meant they could collaborate in real time to write the laboratory report while following protocols—being in the same room but not allowed to have group discussions. Thus, they maintained the designation of a face-to-face class yet safely engaged in group work that embodied Chickering and Gamson's (1987) foundational principles of effective teaching.

The goal of the activities described above is rooted in the principles of emergency preparation. Regardless of the approach utilized, beliefs, technical skills, professional development, training, and accessibility are continually identified as influential factors on teachers' self-efficacy (Canbay, 2020; Kwon et al., 2019). As students worked through each of these technological tools together in the face-to-face environment, technical issues could be dealt with immediately with peer support or flagged for follow-up by the university IT personnel. Importantly, exploring each of these platforms prior to needing them in a stressful, unplanned situation meant that students could think deeply about how to integrate these tools into their own pedagogical practice while supported by their instructor who could share her own TPACK knowledge in real time.

> *Katarin:* Walking the walk and talking the talk was my mantra for the fall of 2020 when I asked the science preservice teachers once they found their "island" to turn on their technology and navigate to the university website. It became very clear at that moment of the range of ability in the class; most were able to Google and find the site; so together we worked our way through a series of tasks. In addition to working through the tasks, I wrote the instructions

on the whiteboard. Why not a PowerPoint? Because you flip pages in a PowerPoint, and I needed the students to stay with me. If they got lost, all the steps were still visible and they could take a picture of the steps. This led to a task of where to put the pictures, again working with the systems. This step-by-step approach, at first, seemed to be over the top; however, later in the course, when they needed to remember "how to," students could go back and refresh their memory and carry on with the task—or use these ideas and methods to have the technology be the tool and not the driver of the lesson.

Wendy: We need to stop perpetuating the myth that the students in our classrooms know how to use technology merely because they grew up with it. They only know what they use, and they may not even fully grasp all the things that piece of software or device can do. And the assumption that this generation all prefers technology is a dangerous perspective to have when designing online learning experiences. In ERT, we had to take the "we're all in this together and we'll muddle through" approach, but we need to also move forward in developing online pedagogical theories and practices.

Wendy was not surprised with the limited technical knowledge of the students; in her work, she realized that most users tended to use only the tools that were necessary and afforded little time to explore the pedagogical possibilities unless coached to do so. Birisci and Kul (2019) revealed that the most important predictor of preservice teachers' self-efficacy beliefs was linked to positive technology experiences, which could be attributed to increased knowledge around technology use. These results highlight how knowledge of technology may be a more influential factor for inservice teachers who have less experience with current available technologies.

3.3.1.5 Demonstrating Pedagogical Prowess

As Guevara et al. (2021) noted, "There is no single approach to providing [an] effective and engaging online [learning experience; … moreover,] the media or platform itself does not foster learning, but instead it is the underpinning instructional method or pedagogy that promotes online learning" (p. 1). Pedagogical prowess—however elusive that capability may be in a rapidly evolving digital context—requires time and space to practice and reflect on each experience and deliberately assess its benefits and challenges.

The preservice teachers were able to put their technological emergency preparation to the test early in the term when Hurricane Teddy hit the province. Based on the Canadian Hurricane Centre's projected track for Teddy, Katarin chose to enact the class's ERT plan.

Katarin: Once the content was covered in Monday morning's class, I reviewed the ERT plan for Wednesday's class. The face-to-face class would be held via Collaborate at its regular time. Students could attend from home or, if the campus remained open, they could come to our classroom. If they lost power or Internet, they were to send a text message to me so that I would know that they were safe and that they could watch the recording later. The point here was that they had the power to choose based on their own context, given the impending weather and the distance that many had to travel. On Wednesday, the university stayed open. 8:00am looked more like 8:00pm; it was as dark as night, the rain was falling in sheets, and the trees at times were bowing as if greeting royalty. I signed on to the Collaborate system from home, hoping that I would not lose my own Internet connection. Only two students were not able to join the live session because of power outages in their rural location. The class, from my perspective, was a success; students were present and engaged, we accomplished everything that was in the outline and a bit more. Later, some students sent emails thanking me for holding the class online as they felt safer staying at home and not driving to campus due to the weather and travel conditions.

Wendy: There is a dichotomy at play on most campuses, where the desire for standard practices and maintaining the traditional face-to-face experience bumps up against a faculty member's desire to test the boundaries of systems and be innovative and responsive to students' learning needs. In my work as instructional designer, faculty [members] have admitted that they were reluctant to try something new in case it failed and their teaching ability was questioned. I have even had teaching colleagues question what I was doing in my own course section, worried that students were comparing how shared learning outcomes were being achieved differently in different sections. How has teaching excellence and academic freedom become a competition of sameness? What about risk-taking and innovation? Why do we assume face-to-face delivery is what is most appropriate for learners and for content and context?

As Katarin's vignette reveals, the advanced planning for ERT worked. She shifted a face-to-face class into the online environment because of a predicted environmental event that undoubtedly would have impacted student attendance in a face-to-face class. Her students were familiar with the various platforms since she had taken time in prior classes to explain the digital tools and to practice using them in a controlled and supportive environment. This class was a successful example of ERT in action, despite the administration's concern that Katarin's proactive actions had put her colleagues at a disadvantage as they had not made similar preparations with their students. This speaks to the need of a unified and collective approach to ERT and pedagogical innovation—one where departmental and administrative supports and permissions are not ambiguous and where students and instructors feel supported in their pedagogical decisions. To address the gaps in training that continue to persist, administrators need to focus not only on providing structural supports to help develop instructors' technological proficiency but also be aware of and work to create the cultural conditions that support both self-directed and collaborative learning (Barton & Dexter, 2020).

Fall practicum went forward with preservice teachers placed in face-to-face classrooms, teaching K–12 students and using a myriad of technology tools and applications. Some experienced temporary school closures due to COVID-19 exposures and had to transition to online learning via the Google suite of applications or by sending home paper packages that students could complete until it was safe to return to in-person learning. With their increased self-efficacy and skill in technological applications (Bandura, 1993; Kwon et al., 2019), they were able to ride not only the COVID-19 waves but also their own learning curve of teaching science. The January 2021 term on campus was an opportunity to reflect on their growing confidence and skills as educators.

Katarin: Isolation protocols meant the first two weeks of winter term classes were online. It was an opportunity for my science education students to be the online instructors, tasked with modelling a resource they used during a lesson from their practicum. Then they reported on how they delivered it to their class, their students' perceptions of it, and what tweaks they would make to the lesson. In these micro-teaching events of 20–30 minutes, the preservice teacher was the moderator, moving their 33 colleagues (the "students") in and out of breakout rooms, sharing screens, creating space for discussion and collaboration in authentic online professional learning community. They did, however, also comment on their nervousness to "run the show" and acknowledged the level of organization needed to manage the number of different informational feeds coming into the instructor (e.g., text, voice, screen-share walk-throughs, PowerPoint, breakout rooms, etc.). Moving to online delivery requires more than putting up a PowerPoint and lecturing; creating a seamless online class is harder than it looks.

Wendy: Teaching that looks effortless requires a lot of effort. This is the case in any grade level and in any modality. The whims of technology and the diversity within classrooms in terms of learner readiness and ability means a Plan B—and sometimes a Plan C—is needed. The preservice teacher education classroom can be a valuable testing ground for trying new techniques while supported by an experienced educator who can help unpack the "why" of the lesson in terms of design and outcomes. Although forced, this pandemic resulted in a valuable experience for these soon-to-be teachers to learn to teach online and to experience the level of management as well as the pedagogical challenges that go alongside teaching online.

Katarin: The remainder of the semester was filled with face-to-face learning during COVID-19 with changing health protocols. As the case numbers dropped, instructors were permitted to use base groups for discussions or activities and students were allowed to talk to each other in person. This was an enormous step forward in that it allowed for the science class to work in base groups to complete their laboratory. With mandatory masking, we could have face-to-face, in-person discussions instead of through Collaborate while conducting experiments. A return to base groups of 3 students each allowed for fewer setups, which meant less preparation time, fewer materials, less expense—instead of 33 individual setups, it was down to 11. I was over the moon! Do you have any idea how much work it is to do 33 individual setups by yourself?

Interestingly, the class still found ways to include more digital components within their laboratory than had been evident in previous science classes over the years, even though the use of technology had always been encouraged. For example, some of the items used and introduced by the preservice teachers included instructions and write-ups completed via MS Forms, YouTube videos for explanations and other resources, Menti meters and Quizlets as formative assessment tools, photos and videos to capture data, various apps as needed for the labs, programming software such as Scratch, etc. It was as if technology was no longer an afterthought or something "posh"—it was just another pencil in the pencil case. The ERT scaffolding that had been put in place at the beginning of the course to be "pulled off the shelf" if and when needed was now being used as part of the daily pedagogy to enhance student learning just as adding colour would to a diagram for understanding and clarity. It was a beautiful thing.

The main takeaway from all conversations and analysis was that students were learning and embedding the technology into their pedagogy, which is something that we, as educational specialists, have been trying to do for years. Most students want interactive and hands-on experiences to enhance their learning and want to work collaboratively. This pandemic made it abundantly clear that there is "value in doing the work and making the decisions for implementing procedures, including troubleshooting" (Jeffery & Bauer, 2020, p. 2484).

3.4 Final Thoughts

Although the COVID-19 pandemic was a surprising disruption for many across the globe, it resulted in important learning to draw upon in the likelihood of future events requiring ERT. With a temporary, quick transition of courses to digital delivery, an inherent challenge is anticipating, accepting, and overcoming the high probability of errors due to strained technological infrastructure, the educator's lack of knowledge and comfort, and the students' anxiety and confusion (Burns et al., 2020; Caron et al., 2022; Miller, 2020). Furthermore, as Donham et al. (2022) noted, "having communication tools and the right technology was critical for both students' and instructor's success during the transition to ERT" (p. 20). Expectations of perfection are unrealistic in the best of times. Educators transitioning to online delivery are bound to make mistakes; but with reflection on the how and why things unfolded the way they did, those mistakes can be beneficial tools for pedagogical growth (Donham et al., 2022; Kraglund-Gauthier & Moseley, 2019; Miller, 2020). These elements of pedagogical growth can strengthen the moments of transition of the class from face-to-face to online whether scheduled or on short notice due to ERT.

Blended learning is much more than a temporary fix for ERT; rather, it exhibits benefits, challenges, and outcomes even though the boundaries of each may seem blurred. It requires a fundamental shift in focus from teaching to learning (Donham et al., 2022; Dziuban et al., 2018). Pedagogical lessons from teacher education reveal how deep learning can result in a shift from lecture to student-centred instruction in which students become active and interactive learners throughout the entire course, including in-person classes (Andrade & Coutinho, 2017; Donham et al., 2022; Dziuban et al., 2018; Kraglund-Gauthier, 2020; MacLeod & Kraglund-Gauthier, 2015). Thus, an unexpected outcome of ERT has been not only heightened recognition of the impacts of socio-economic, racial, and cultural on student learning (McKenzie, 2021), but also an increased awareness of and focus on the importance of pedagogy in higher education regardless of modality to increase student engagement, motivation, and productivity (Caron et al., 2022).

The findings from this self-study of ERT and the pedagogical implications of a rapid transition to a blended and fully online classroom echo and push the boundaries of previous research in online pedagogy (Jeffery & Bauer, 2020; Kraglund-Gauthier & Moseley, 2019). As shown in this chapter, a pedagogical approach rooted in TPACK combines socialization opportunities of the classroom where "'shoulder-to-shoulder' [learning] … provided them the ability to check-in with each other as instruction rolled out, supporting and encouraging motivation to engage and persist" (Jeffery & Bauer, 2020, p. 2482) with the technologically enhanced, active learning possibilities of the online environment (Dziuban et al., 2018; Lai et al., 2016).

It is crucial for instructors to "think critically about the process of learning and the type and quality of assessment when making decisions about what tools and resources to use in a face–to-face learning environment that can also be applied to learning in the online context" (MacLeod et al., 2022). As Beason-Abmayr et al. (2021) discovered in their analysis of student learning outcomes, some aspects of course learning did not transfer as well, such as modifications to in-class, hand-on learning activities and replicating the engagement in small group discussions when students were reticent to use their audio or video tools. Institutions' motivations for implementing a blended learning model, such as enhanced learning and greater flexibility for students as well as part of emergency preparedness plans, need to be clearly communicated to instructors and students (Caron et al., 2022; McKenzie, 2021; Owston & York, 2018). By clearly explaining the *why*, leaders help instructors and students understand the significance of resiliency in learning and become key supporters of the initiative.

From the vignettes discussed above, it is clear that the education system has come to a fork in the road. Predicated by the adage "you can't bloom until you Maslow" (McKenzie, 2021), the question then becomes how best meet students' basic learning needs and expectations for online course delivery (Caron et al., 2022; Garris & Fleck, 2022). To do so requires forging a path where the transitioning between pedagogical modalities becomes ever more fluid; that is, the devices and tools are no longer siloed into online learning, blended learning, or face-to-face learning, but rather connect in and with learning regardless of the environment or platform or modality, but also take into account students' resources for learning online (Kajjimu et al., 2023). Such a transformation will result in the global society being able to seamlessly transition whatever crisis may be at hand as seen in the examples discussed within this chapter.

References

Abrami, P. C., Bernard, R. M., Bures, E. M., Borokhovski, E., & Tamim, R. M. (2011). Interaction in distance education and online learning: Using evidence and theory to improve practice. *Journal of Computing in Higher Education, 23*(2–3), 82–103. https://doi.org/10.1007/s12528-011-9043-x

Akcil, U., Uzunboylu, H., & Kinik, E. (2021). Integration of technology to learning-teaching processes and Google Workspace tools: A literature review. *Sustainability, 13*(9), 5018. https://doi.org/10.3390/su13095018

Alhassan, R. (2017). Exploring the relationship between Web 2.0 tools self-efficacy and teacher's use of these tools in their teaching. *Journal of Education and Learning, 6*(4), 217–228.

Allen, I. E., & Seaman, J. (with Lederman, D., & Jaschik, S.). (2012). *Conflicted: Faculty and online education, 2012. Inside Higher Ed* & Babson Survey Research Group. http://www.insidehighered.com/sites/default/server_files/files/IHE-BSRG-Conflict.pdf

Andrade, M., & Coutinho, C. (2017). Implementing flipped classroom in blended learning environments: A proposal based on the cognitive flexibility theory. *Journal of Interactive Learning Research, 28*(2), 109–126. https://www.learntechlib.org/primary/p/174184/

Bandura, A. (1993). Perceived self-efficacy in cognitive development and functioning. *Educational Psychologist, 28*(2), 117–148. https://doi.org/10.1207/s15326985ep2802_3

Barton, E. A., & Dexter, S. (2020). Sources of teachers' self-efficacy for technology integration from formal, informal, and independent professional learning. *Educational Technology, Research and Development, 68*(1), 89–108. https://doi.org/10.1007/s11423-019-09671-6

Bates, A. W. (2022). *Teaching in a digital age: Guidelines for designing teaching and learning* (3rd ed.). Tony Bates Associates Ltd. https://pressbooks.bccampus.ca/teachinginadigitalagev3m/

Beason-Abmayr, B., Caprette, D. R., & Gopalan, C. (2021). Flipped teaching eased the transition from face-to-face teaching to online instruction during the COVID-19 pandemic. *Advances in Physiological Education, 45*, 384–389. https://doi.org/10.1152/advan.00248.2020

Birisci, S., & Kul, E. (2019). Predictors of technology integration self-efficacy beliefs of preservice teachers. *Contemporary Educational Technology, 10*(1), 75–93. https://doi.org/10.30935/cet.512537

Boettcher, J. V., & Conrad, R. M. (2021). *The online teaching survival guide: Simple and practical pedagogical tips.* (3rd ed.). Jossey-Bass.

Buksh, S. (2020). Maslow before Bloom. *Independent. Education, 50*(2), 20–21.

Burns, A., Danyluk, P., Kapoyannis, T., & Kendrick, A. (2020). Leading the pandemic practicum: One teacher education response to the COVID-19 crisis. *International Journal of e-Learning and Distance Education, 35*(2). https://files.eric.ed.gov/fulltext/EJ1299367.pdf

Canbay, F. (2020). EFL teachers' views about technology integration in English language teaching: A case study. *i-manager's Journal on English Language Teaching, 10*(2), 54–63. https://doi.org/10.26634/jelt.10.2.16425

Caron, E. E., Drody, A. C., Hicks, L. J., & Smilek, D. (2022). The impact of a global pandemic on undergraduate learning experiences: One year later. *Trends in Neuroscience and Education, 29*, 100184. https://doi.org/10.1016/j.tine.2022.100184

Chickering, A. W., & Gamson, Z. F. (1987). Seven principles for good practice in undergraduate education. *AAHE Bulletin, 39*(7), 3–7. https://files.eric.ed.gov/fulltext/ED282491.pdf

Columbia University Center for Teaching and Learning. (2020). *Adapting your face-to-face course to a fully online course: A guide.* Columbia University CTL. https://ctl.columbia.edu/resources-and-technology/teaching-with-technology/teaching-online/adapting-your-course/

Crawford, K. (1996). Vygotskian approaches in human development in the information era. *Educational Studies in Mathematics, 31*(1/2), 43–62. https://doi.org/10.1007/BF00143926

Donham, C., Barron, H. A., Alkhouri, J. S., Kumarath, M. C., Alejandro, W., Menke, E., & Kranzfelder, P. (2022). I will teach you here or there, I will try to teach you anywhere: Perceived supports and barriers for emergency remote teaching during the COVID-19 pandemic. *International Journal of STEM Education, 9*(19), https://stemeducationjournal.springeropen.com/articles/10.1186/s40594-022-00335-1#change-history

Dziuban, C., Graham, C. R., Moskal, P. D., Norberg, A., & Sicilia, N. (2018). Blended learning: The new normal and emerging technologies. *International Journal of Educational Technology in Higher Education, 15*(1), Article 3. https://doi.org/10.1186/s41239-017-0087-5

Efwinda, S., & Mannan, M. N. (2021). Technological pedagogical and content knowledge (TPACK) of prospective physics teachers in distance learning: Self-perception and video observation. *Journal of Physics: Conference Series, 1806*(1). https://doi.org/10.1088/1742-6596/1806/1/012040

Esfijani, A., & Zamani, B. E. (2020). Factors influencing teachers' utilization of ICT: The role of in-service training courses and access. *Research in Learning Technology,* 28. https://doi.org/10.25304/rlt.v28.2313

Friesen, N. (2011). *The place of the classroom and the space of the screen: Relational pedagogy and Internet technology.* Peter Lang.

Garris, C. P., & Fleck, B. (2022). Student evaluations of transitioned-online courses during the COVID-19 pandemic. *Scholarship of Teaching and Learning in Psychology, 8*(2), 119–139. https://doi.org/10.1037/stl0000229

Glaser, B. G. (1965). The constant comparative method of qualitative analysis. *Social Problems, 12*(4), 436–445. https://doi.org/10.2307/798843

Government of Nova Scotia. (2020). *Coronavirus (COVID-19): Alerts, news and data.* https://novascotia.ca/coronavirus/alerts-notices/

Guevara, K., Fattah, L., Ritt-Olson, A., Yin, P.-L., Litman, L., Farouk, S. S, O'Rourke, R., & Mayer, R. E. (2021). Busting myths in online education: Faculty examples from the field. *Journal of Clinical & Translational Science, 5*(1), Article e149. https://doi.org/10.1017/cts.2021.808

Hodges, C. B., Moore, S., Lockee, B. B., Trust, T., & Bond, M. A. (2020). *The difference between emergency remote teaching and online learning*. EDUCAUSE Review. https://er.educause.edu/articles/2020/3/the-difference-between-emergency-remote-teaching-and-online-learning

Hu, D., Yuan, B., Luo, J., & Wang, M. (2021). A review of empirical research on ICT applications in teacher professional development and teaching practice. *Knowledge Management & E-Learning, 13*(1), 1–20. https://doi.org/10.34105/j.kmel.2021.13.001

Hunter, R. (2004). *Madeline Hunter's mastery teaching: Increasing instructional effectiveness in elementary and secondary schools* (2nd ed.). Corwin Press.

Jeffery, K. A., & Bauer, C. F. (2020). Students' responses to emergency remote online teaching reveal critical factors for all teaching. *Journal of Chemical Education, 97*(9), 2472–2485. https://doi.org/10.1021/acs.jchemed.0c00736

Kajjimu, J., Dreifuss, H., Tagg, A., Dreifuss, B., & Bongomin, F. (2023). Undergraduate learning in the COVID-19 pandemic: Lessons learned and ways forward. *Advances in Medical Education and Practice, 14*, 355–361. https://www.ncbi.nlm.nih.gov/pmc/articles/PMC10095942/pdf/amep-14-355.pdf

Koehler, M. J., & Mishra, P. (2009). What is technological pedagogical content knowledge? *Contemporary Issues in Technology & Teacher Education, 9*(1), 60–70. http://www.citejournal.org/vol9/iss1/general/article1.cfm

Kraglund-Gauthier, W. L. (2020). Using participatory action research to support pedagogical processes in postsecondary online and blended spaces. In G. Parchoma, M. Power, & J. Lock (Eds.), *The finest blend: Graduate education in Canada*. Athabasca University Press. https://doi.org/10.15215/aupress/9781771992770.01

Kraglund-Gauthier, W. L., & Moseley, J. (2019). Building teaching–learning capacities of online nurse educators: Using TPACK to frame pedagogical processes and identify required supports. *Canadian Journal of Learning and Technology, 45*(1), 1–21. https://files.eric.ed.gov/fulltext/EJ1214588.pdf

Kwon, K., Ottenbreit-Leftwich, A. T., Sari, A. R., Khlaif, Z., Zhu, M., Nadir, H., & Gok, F. (2019). Teachers' self-efficacy matters: Exploring the integration of mobile computing device in middle schools. *TechTrends, 63*(6), 682–692. https://doi.org/10.1007/s11528-019-00402-5

Lai, M., Lam, K. M., & Lim, C. P. (2016). Design principles for the blend in blended learning: A collective case study. *Teaching in Higher Education, 21*(6), 716–729. https://doi.org/10.1080/13562517.2016.1183611

Leach, J., & Moon, B. (2008). *The power of pedagogy*. Sage.

Leavy, P. (2023). *Research design: Quantitative, qualitative, mixed methods, arts-based, and community-based participatory research* approaches (2nd ed.). The Guilford Press.

Loughran, J. (2007). Researching teacher education practices: Responding to the challenges, demands, and expectations of self-study. *Journal of Teacher Education, 58*(1), 12–20. https://doi.org/10.1177/0022487106296217

MacLeod, K. A., & Kraglund-Gauthier, W. L. (2015). A case study of infusing technology into pre-service secondary science teacher learning: Conceptions and attitudes while navigating changing digital landscapes. *Journal of European Education, 5*(2), 1–15. https://doi.org/10.18656/jee.24828

MacLeod, K. A., Kraglund-Gauthier, W. L., & Griffiths, G. (2022). The efficacy of online games for teaching ocean literacy in Nova Scotia: Results from a pilot study. In M. Jacobsen & C. Smith (Eds.), *Online learning and teaching from kindergarten to graduate school* (pp. 526–553). Canadian Association of Teacher Education. https://doi.org/10.11575/PRISM/40509

Merriam, S. B. (1998). *Qualitative research and case study applications in education*. Jossey-Bass.

McKenzie, L. (2021). *Back on track: Helping students recover from COVID-19 learning disruption*. Inside Higher Ed. https://www.insidehighered.com/content/back-track-helping-students-recover-covid-19-learning-disruption

Miller, J. A. (2020, December 7). Eight steps for a smoother transition to online teaching. *Faculty Focus*. https://www.facultyfocus.com/articles/online-education/online-course-design-and-preparation/eight-steps-for-a-smoother-transition-to-online-teaching/

Moore, M. G. (1989). Editorial: Three types of interaction. *American Journal of Distance Education, 3*(2), 1–7. https://doi.org/10.1080/08923648909526659

Osterman, K. F., & Kottkamp, R. B. (1993). *Reflective practice for educators: Improving schooling through professional development.* Corwin Press.

Owston, R., & York, D. (2018). The nagging question when designing blended courses: Does the proportion of time devoted to online activities matter? *The Internet & Higher Education, 36*, 22–32. https://doi.org/10.1016/j.iheduc.2017.09.001

Pandey, S. K., & Mishra, M. (2021). Interaction in distance education and online learning: Using evidence and theory to improve practice. *TechnoLEARN, 11*(1), 1–8. https://doi.org/10.30954/2231-4105.01.2021.1

Public Safety Canada. (2019). *Emergency management strategy for Canada: Toward a resilient 2030.* https://www.publicsafety.gc.ca/cnt/rsrcs/pblctns/mrgncy-mngmnt-strtgy/index-en.aspx

Public Safety Canada. (2022, September 20). *Emergency management.* https://www.publicsafety.gc.ca/cnt/mrgnc-mngmnt/index-en.aspx

Rodríguez-Muñiz, L. J., Burón, D., Aguilar-González, Á., & Muñiz-Rodríguez, L. (2021). Secondary mathematics teachers' perception of their readiness for emergency remote teaching during the COVID-19 pandemic: A case study. *Education Sciences, 11*(5), 228. https://doi.org/10.3390/educsci11050228

Rogers, P. L. (2011). *Designing instruction for technology-enhanced learning.* IGI Global.

Shulman, L. S. (1986). Those who understand: Knowledge growth in teaching. *Educational Researcher, 15*(2), 4–14. https://doi.org/10.3102/0013189X015002004

Skilling, K., & Stylianides, G. J. (2020). Using vignettes in educational research: A framework for vignette construction. *International Journal of Research & Method in Education, 43*(5), 541–556. https://doi.org/10.1080/1743727X.2019.1704243

Torres, S. (2009). Vignette methodology and culture-relevance: Lessons learned through a project on successful aging with Iranian immigrants to Sweden. *Journal of Cross-Cultural Gerontology, 24*(1), 93–114. https://doi.org/10.1007/s10823-009-9095-9

Vygotsky, L. S. (1978). *Mind in society: The development of higher psychological processes.* Harvard University Press.

Wenger, E. (2006). *Communities of practice: A brief introduction.* https://scholarsbank.uoregon.edu/xmlui/bitstream/handle/1794/11736/A%20brief%20introduction%20to%20CoP.pdf

World Health Organization. (2020, June 29). *Archived: WHO timeline- COVID-19.* WHO. https://www.who.int/news/item/27-04-2020-who-timeline---covid-19

Yin, R. K. (2018). *Case study research and applications: Design and methods* (6th ed.). Sage.

Chapter 4
Teaching Science to English Language Learners in an Urban Canadian Context: Secondary Teachers' Perspectives

Latika Raisinghani

4.1 Introduction

Although immigrant students in Canada are reported to have better educational achievements than immigrant students in other OECD (Organisation for Economic Co-operation & Development) countries, many English Language Learners (ELLs) continue to experience academic difficulties (Cummins et al., 2012). This chapter stems from a larger qualitative case study of K–12 teachers' perspectives on the viability of culturally responsive teaching in science and mathematics classrooms in urban western Canada (Raisinghani, 2018). Here, I focus on the secondary science teachers and explore their perspectives on teaching ELLs.

Characterized as an "ethnocultural mosaic" (Chui & Flanders, 2013, p. 6), Canada is one of the highest immigrant-receiving countries of the world (Joshee et al., 2016; Milford & Tippett, 2016). In the first Canadian census in 1871, only 20 countries of origin were reported; most of the population stated either British (60.5%) or French (31.1%) origin (Vézina & Maheux, 2017). The most recent 2021 census indicated that Canada has people of 450 ethnic origins (Statistics Canada, 2022). Thus, cultural diversity is "a fact of life in Canada" (Gill & Chalmers, 2007, p. 552).

A large proportion of Canada's immigrant population is concentrated in the nation's urban centres: Toronto, Montreal, and Vancouver (Joshee et al., 2016). This study was conducted in the Vancouver School District of British Columbia (BC) in which nearly 50% students speak a language other than English at home (Vancouver School Board, 2024).

Responding to the richness of cultural diversity of its population, Canada is the first country that established a national multicultural policy in 1971 (Ghosh & Galczynski, 2014; Moodley, 1995). Since the inception of this federal policy, education has been

L. Raisinghani (✉)
University of Winnipeg, Winnipeg, MB, Canada
e-mail: l.raisinghani@uwinnipeg.ca

C. D. Tippett and T. M. Milford (eds.), *Seeing Science Through the Eyes of Canadian Teachers and Learners*, Contemporary Trends and Issues in Science Education 54,
https://doi.org/10.1007/978-3-031-88785-7_4

considered as the key to addressing issues pertaining to cultural diversity (Joshee et al., 2016). However, rather than having a federal education system, education in Canada is primarily a provincial responsibility (Milford & Tippett, 2016). Ministries or departments of education in each of the ten provinces and three territories are responsible for the establishment of local policies as well as for overseeing their execution in coordination with regional school districts. Intergovernmental cooperation is illustrated by the collaboration of these education ministries on pan-Canadian educational priorities under the Council of Ministers of Education, Canada (CMEC), which is an intergovernmental body founded to discuss policy issues and undertake educational initiatives of national concern and to represent the country internationally. In 1997, CMEC introduced a *Common Framework of Science Learning Outcomes, K to 12*, which informs curriculum development and the delineation of critical scientific literacy among students in various Canadian jurisdictions. While this *Common Framework* acknowledges that "personal and cultural preconceptions and prior knowledge … [affect student learning and encourages students to share developed understandings] with others using everyday language, in diverse situations that respect a wide variety of learners" (CMEC, 1997, p. 7), it does not explicitly endorse multicultural and multilingual discourses or provide any guidelines that could support culturally diverse students' learning in science classrooms.

4.1.1 Culture and Culturally Responsive Teaching

Culture is an encompassing and dynamic process that can be informed by ways of knowing, languages, actions, attitudes, beliefs, values, and traditions that are unconsciously or consciously acquired by members of a specific community (Banks, 2009; Thomas, 2003). Similarly, cultural diversity goes beyond differences in ethnicity, language, gender, religion, socioeconomic status, age, or dis/ability; rather, it is based on differences across all cultural aspects (Horowitz et al., 2005; Lee, 2010; Raisinghani, 2018, 2019).

Recognizing cultural diversity as a strength, culturally responsive teaching (CRT) emphasizes that culture plays an important role in how students receive and interpret knowledge (Gay, 2010). With a CRT approach to meet the educational needs of diverse students, it is essential for teachers to strengthen their own cultural knowledge base by recognizing students' "funds of knowledge" (Nashon & Anderson, 2013, p. 403). Highlighting the establishment of welcoming, safe, and inclusive classroom environments that are conducive to learning of all students, CRT encourages cultural caring where teachers understand the cultural and learning preferences of their diverse students and design learning environments that do not hinder academic effort and achievement (Gay, 2010). CRT requires teachers to understand that culture and communication are mutually interdependent and that their interaction plays a key role in improving cross-cultural communications. It encourages teachers to acquire cultural competence by critically examining their own cultural biases via continually questioning sociocultural inequities and codes of power that may influence pedagogical relationships in cultural diversity-rich classrooms (Delpit, 2006; Ladson-Billings, 2001). CRT invites teachers to embrace culturally responsive curricula by analyzing and evaluating formal school curricula and how it is presented by contextualizing it with/in race, ethnicity, gender, socioeconomic class, and other sociocultural variables to respond to the complexity of cultural diversity (Henry, 2017).

Unfortunately, little recognition has been given to the linguistic and cultural resources that students bring to science classrooms; and teachers often find it difficult to integrate students' cultural ways of knowing in subjects such as science (Gay, 2010; Lee, 2005). Few teachers consider science acultural (Krugly-Smolska, 2013); but many feel ill-equipped to meet the needs of culturally diverse students, expressing concern about the content of their teacher education programs (Egbo, 2019; Raisinghani, 2018; Teel & Obidah, 2008). Preparing science teachers who can effectively support ELLs in secondary classrooms is a critical issue in science education (Lyon et al., 2018). Often teachers' perspectives of language and teaching are prejudiced within subject-specific boundaries, and they may view efforts to promote language learning to support ELLs within science classrooms as detrimental to the image of science (Arkoudis, 2003). Development of teacher education and professional development programs that promote knowledge of science content and pedagogy (Olson et al., 2015), while cultivating the cultural competence that may equip teachers in supporting science learning of ELLs (Oliver, 2017), is a pressing issue in multicultural societies such as Canada.

4.1.2 Theoretical Framework

A theoretical framework of (trans-multi)culturally responsive education (Raisinghani, 2018; Fig. 4.1) may provide insights to support teachers in developing cultural competency. Informed by Goodenough's (1976) notion of multiculturalism as a normal human experience, this framework emphasizes that every individual holds multiple cultural identities; therefore, teachers should appreciate the multiplicity of their students' cultural identities. The framework comprises four quadrants that collectively represent key elements of (trans-multi)culturally responsive education. The upper-right quadrant represents Nieto's (2000) critical multicultural education perspective and reflects four levels of multicultural education that inform critical understandings of difference as a continuum: tolerance (lowest level); acceptance; respect; and affirmation, solidarity, and critique (highest level). Hence, a critical multicultural education advocates that teachers, while engaging with students from diverse cultural backgrounds, should move beyond merely tolerating, accepting, or respecting and engage with the difference with solidarity and critique (Nieto, 2000). The lower-right quadrant embeds Delpit's (2006) cultural codes of power with five key elements of Gay's (2010) notion of CRT (i.e., cultural knowledge base of teachers, cultural relevance of curricula, cultural caring and building a learning community, cultural communication, and cultural congruency in classroom instruction) that are critical to ensure welcoming learning experiences for diverse students.

The lower-left quadrant, representing key aspects of transformational multicultural education (Keating, 2007), provides insights that may help teachers in embracing nonbinary epistemological and pedagogical methods by engaging students in transcultural dialogues to experience nonbinary forms of opposition, resistance, and transformation and to invite relational and holistic connectedness in their learning process. The upper-left quadrant represents various dimensions of multicultural education (Banks & Banks, 2010). These dimensions call for a systemic change as they reflect that content integration and knowledge construction are just the beginning steps and that a holistic multicultural education demands prejudice reduction, equity pedagogy, and an empowering school culture where all students feel belonged and empowered to contribute to their own and peers' learning.

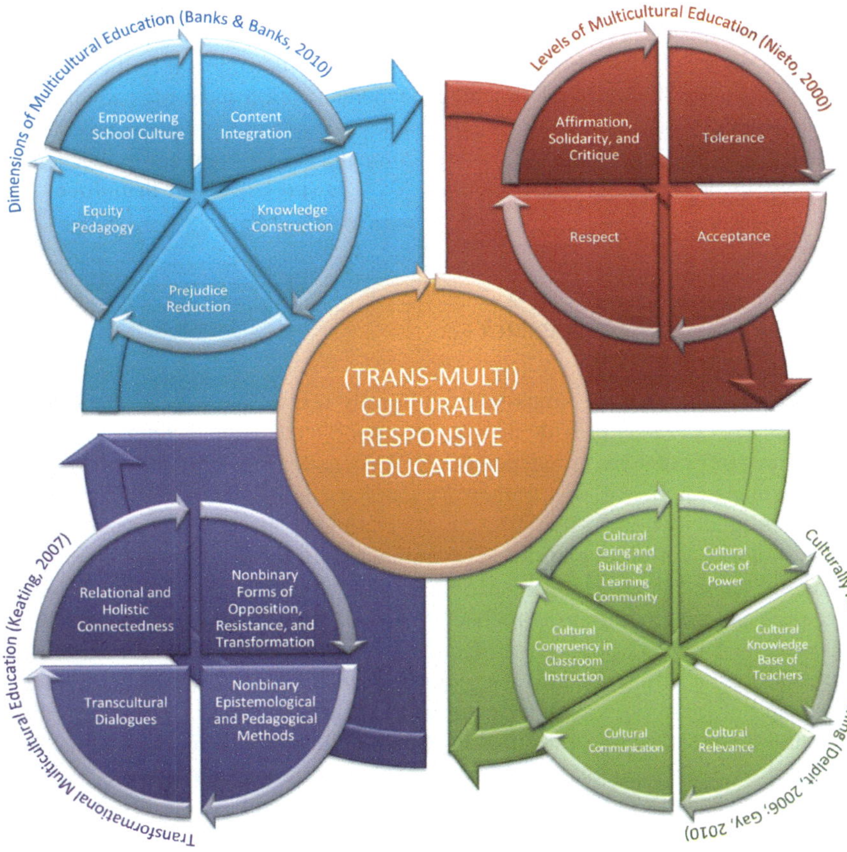

Fig. 4.1 (Trans-multi)culturally responsive education framework (Raisinghani, 2018, p. 71)

Positioning culture as a dynamic way of life and cultural diversity as all lived experiences that students may bring to schools, a (trans-multi)culturally responsive education allows teachers to strategically plan instruction and assessment with a strength-focused perspective beginning with students' diverse prior knowledge and skills. Students' lived experiences and cultural ways of knowing are validated, and teaching becomes holistic. Translated into science classrooms, a (trans-multi)culturally responsive education framework emphasizes undoing and unlearning of abstract, acultural teaching of science and adopting culturally responsive pedagogies. This framework served as a guiding lens for my research as I explored the complexities of teaching science in culturally diverse urban secondary classrooms.

4.1.3 English Language Learners

My research was concerned with ELLs and their linguistic diversity, one of many aspects of cultural diversity. As Cummins et al. (2012) noted, the development of literacy-related concepts and skills in home language (L1) and dominant language (L2) are interdependent. Many ELLs struggle academically because they are developing general English language skills while attempting to develop proficiency in complex academic English. Research suggests that ELLs may learn conversational English or basic interpersonal communicative skills (BICS) in one to two years, but it may take five or more years to develop cognitive academic language proficiency (CALP), which is the academic English proficiency comparable to their English L1 peers (Cummins et al., 2012).

While learning science involves concrete experiences, it is largely a language-based discipline that requires a wide range of skills in both general and academic content-based language (Oliver, 2017). The linguistic demands of science present challenges for ELLs because they must learn the language of science while they are still learning English (Reid et al., 2019). In secondary science classrooms, ELLs can be deprived of rich learning experiences (e.g., engaging in inquiry, participating in design-based tasks, and implementing projects that require active meaning making) because often the instruction for ELLs "tends to focus either on the basics, or on the development of academic language separate from the content" (Callaham et al., 2019, p. 184).

Indeed, teachers need to recognize that there is a complex interplay among science learning, literacy development, and English language proficiency (Lee, 2005). Learning science involves developing inquiry-oriented thinking, scientific reasoning, and evidence-based communication skills that require learning lexical and grammatical structure of the language of science (Cardozo-Gaibisso et al., 2017; Kelly et al., 2023). Therefore, to help ELLs learn and effectively communicate in science, teachers need to adopt CRT (Gay, 2010), which could validate students' lived experiences and create opportunities for them to utilize culturally based linguistic resources in science classrooms (Oliver, 2017).

4.2 Methods

To explore the issues that teachers face while teaching science to ELLs in culturally diverse secondary classrooms, I conducted a qualitative case study (Stake, 2000) by utilizing phenomenographic methods (Marton, 2014). I was guided by the epistemological belief that everyone experiences the world differently and an individual's experience of any phenomenon is just one of the ways in which that particular phenomenon could be experienced (Marton, 1986). Each participating teacher may have experienced the phenomena of teaching science to ELLs differently within the context of their classroom. Hence, a careful account of each teacher's ways of experiencing the phenomenon could enhance understanding of teaching science to ELLs.

Through purposeful sampling (Merriam, 1998), I recruited four secondary science teachers from a high school in one of the most culturally diverse school districts in the province (Vancouver School Board, 2024). Two teachers self-identified as White Canadian females, one as a Canada-born Indo-Canadian female, and the other as a Chinese Canadian male who immigrated to Canada in the early 2000s. All four teachers had earned a bachelor's degree in education and were certified to teach in BC. In their various science courses, more than half of the students spoke English as an additional language and were designated as ELLs. One teacher was teaching a course called English as a Second Language (ESL) Science with only ELLs enrolled.

This small sample of teachers was not representative of the totality of teachers in the wider Canadian context. However, as Stake (2000) mentioned, the focus of a case study approach is not to defend the typicality of the case but is to decipher "what specifically can be learned from the case" (p. 436). Hence, I aimed to generate thick descriptions of select teachers' perspectives (Merriam, 1998) that would facilitate enhanced understanding of teaching science to ELLs, whether in Canada or wider international contexts.

I collected data through two individual interviews with each teacher; the interviews were complemented with at least four observations of their classroom to understand the contextual details within which they taught. Audio-recorded interviews were transcribed verbatim to generate thick descriptions that were supported by fieldnotes taken during the classroom observations. The (trans-multi)culturally responsive education framework served as a guiding lens for my interpretation of the collected data. I identified emergent themes by looking for similarities and differences in teachers' perspectives in a process of thematic analysis (Miles & Huberman, 1994).

4.3 Findings

In this section, I focus on five key themes that emerged from my analysis and share the four participating teachers' perspectives about teaching science to ELLs. The key themes indicate that low English language competency imposed a cultural barrier, impacted ELLs' participation, resulted in increased behavioural issues, influenced teachers' choice of teaching practices and resources, and posed system-based challenges. In the following sections, I elaborate on each theme and provide example teacher quotes from the interviews or observations as evidence.

4.3.1 Language as a Cultural Barrier

English was the language of instruction for the participating teachers; they recognized that low English language competency could pose a cultural barrier, affecting ELLs' social and academic interactions. These teachers were concerned that ELLs often formed cohesive cultural bubbles and avoided communicating with English-speaking peers. They claimed that many ELLs were not interested in learning English because they could easily communicate with their L1 peers. To avoid cultural bubbles, they often had to ensure that all students were using English to communicate in science. The following interview excerpt conveys these collective concerns:

> There is such a high population in our school of … kids who don't speak English…. The challenges [of students' cultural diversity] are the fact that you do get students feeling left out. And this is where, you know as I was saying, acting like the language police is the only thing that I can do in [Chemistry 11]…. To say that you know that English is our common language and that's the language [I'd] like us all to speak while we are in [the science classroom] and when we are doing work together.… When they [ELLs] get here, everywhere they see someone who speaks the language that they speak so really the impetus to speak English is very low. It [level of English competency] does make it very difficult [to teach science] you know. I do get students who I can see by their writing that have the right idea but they just don't have the words to express it and the higher up you get in school like the senior grades you really do have to be able to express your ideas in a way that are [*sic*] understandable, so it becomes more difficult as they [students] get older to do well in a course like Chemistry if they don't have a good grasp of the [English] language. (Teacher 1)

4.3.2 Differences in ELLs' Participation

The teachers noticed that newly arrived students from non-English speaking countries often found it difficult to navigate engaging in science as inquiry. They stated that ELLs were often hesitant to ask or respond to questions and did not participate in whole-class discussions or other peer interactions and activities, which these teachers had identified as crucial for student success in their classroom (Raisinghani, 2018). They reasoned that ELLs' lack of participation could be due to the students' low

English competency level and prior experiences in the traditional teacher-centric education system of their home country.

> The other thing that I find with students who don't have strong English skills is that they are very reluctant to participate in verbal discussions.... I'll ask a question, and I know that 90% of the class knows the answer, but they don't want to say that aloud. (Teacher 1)

The teachers felt that the differences in ELLs' prior learning experiences in their home country resulted in different expectations regarding learning science and a differing understanding of participation in science classrooms:

> Students have come to Canada to receive a Canadian education, which is often very different from the education system they are used to.... I gave a problem-based lab[oratory] in my Biology class and some of the students were really lost about how to create an inquiry question.... Some of them are new to the Canadian education [system]. (Teacher 2)

4.3.3 Behavioural Issues

The teachers remarked that along with the low level of English competency the number of behavioural issues were arising. An example of a behavioural issue that occurred during my observation of Teacher 1's classroom follows:

> As the class began, the teacher walked around with a basket and asked students to turn in their cell phones and gave them a worksheet to read and begin working on their assigned lab[oratory] activity (of testing pH of given solutions) in pairs or groups of three students (students self-made their pair or groups; no specific instructions about partnering were given). One of the students refused to hand in their phone and just put the worksheet on one side. When the teacher asked this student to put their phone aside and do the assigned experiment, the student just mumbled, 'I don't know.' The teacher then asked the student to pair up with another student or join any of the student groups, but the student just looked down and continued using the phone. The teacher then left the student alone and started giving directions to other student groups. During the next 10–15 min, the teacher went around the room reminding some student groups to speak in English and tried to encourage the student with the cell phone to either work alone or join any of the student groups. But the student just continued using the cell phone with their head bent down. After a while, the teacher picked a file from the teacher's desk and asked this student to take it to the principal's office. The student left the classroom with the cell phone and file and did not return until the class period ended. (Fieldnotes)

After the class, Teacher 1 shared that this student had recently arrived from a non-English speaking country and that no other student in the class spoke the same L1:

> I sent this new student to the principal's office for some errand because I did not want the student to create disciplinary issues by not participating in the lab[oratory] activity and continuing cell phone use in the class.... I have talked with the student many times, but I also have 29 other students in my class. (Fieldnotes)

Another behavioural issue was that often ELLs did not come to class on time:

The schools are gated [in students' home countries] which means [the students] can't leave the school and the classroom, and it's the teachers who rotate and [the students] stay in the classroom. So, it's really a different system whereas here the kids are coming and going [in the classrooms], and I do suspect that the kids start taking the advantage. They usually come ten minutes or so late to the class. (Teacher 3)

4.3.4 Choices of Teaching Practices and Resources

The teachers said that they preferred science textbooks that were two or three decades old because newer textbooks were more text heavy, making them inappropriate in their classroom with many ELLs. Moreover, the teachers claimed that new textbooks did not address issues of cultural diversity:

I don't see that the textbooks are really addressing the issues of cultural diversityCultural diversity.... I think that the authors of the textbooks were thinking that they were teaching a certain uniform group of students ... maybe Canadian students, Canada-born students, I don't know ... It's just, I don't think it is something that is explicitly considered when they [textbooks] are written. (Teacher 2)

The teachers reported utilizing visuals, teaching specific science vocabulary, and modifying their teaching strategies to facilitate understanding of science for diverse students, especially for those who were designated ELLs:

So, just asking a question or trying to start a discussion doesn't really work very well when you have a class with a large number of students who don't speak English. So, rather than just throwing out a question and asking them just on the spot answer ... it's just like you would say to them that I am going to ask you this, discuss it with your partner, and then we will answer it or you would even say that I am going to let you discuss it with your partner and, as these students are discussing it, you might go to this [ELL] and say I am going to ask you this question. So, then they know that it's coming, and they get time to formulate the answer.... I think those things are helpful regardless of if you are an English speaker or not. (Teacher 1)

In the same fashion, the teachers attempted to diversify their instructional and assessment strategies:

I am trying to use a lot of different methods. I am using PowerPoints, but I also use demos, labs that are more hands-on. I try to allow time for them [ELLs] to translate work into their own language before we even get into it. So, if I gave them a lab [activity], I will get them to write, I will give them all the steps [showed a laboratory report handout] but there is a chunk of space that [is] really just to translate those words.... When they are not going to be verbal, I have to find other ways of engagement. (Teacher 3)

The teachers reported altering their expectations for student assignments:

I try to make language a very small aspect of their learning in physics.... Our school is highly academic, and there is a lot of pressure on students to perform better in science. I know this pressure is more for some students ... most of my ELLs [from specific countries] have tutoring classes after school. I tend not to give any assignments that require a lot of writing. Because I can see a lot of difference in the work when these students write in my [Physics 11 and 12] classes and when they bring their written assignments. So, I usually

> give them a lab[oratory] report format and give them time to work on it during class time. (Teacher 4)

Similarly, the teachers tried to simplify the language of the science tests and quizzes they gave:

> When I make up the test questions, [I] really try to keep the language very basic because I'll find that I'll ask a question but the part of the question that's not even the part that is testing your knowledge [creates confusion or difficulty because] ... they [ELLs] don't know the words [in English]. Like for example, I may have a question written as 'Crossing over would occur ...' but then I end up with a whole bunch of hands [with students asking]: 'What do you mean by "occur"?' So, then I changed the question to 'Crossing over happens when ....' (Teacher 1)

4.3.5 System-Based Challenges in Supporting ELLs

Despite these efforts to accommodate ELLs, the teachers said that some system-based challenges must be addressed in order to fully support ELLs' learning of science. They felt that they were not professionally prepared to support culturally diverse students with varying levels of English competency in their science classroom. In addition, they described streaming as a hindrance for ELLs.

First, none of the four participants reported receiving specific teacher education about supporting ELLs in science classrooms. One participant had attended a 1-day professional development program that centered on educational systems in Asia. In lieu of adequate professional training, they felt that they were left on their own and struggled to design science lessons that could meet the needs of their linguistically diverse students. Teacher 3, who despite having no training was assigned to teach ESL Science, described the situation:

> I think it [teaching science to ELLs] has been very challenging for me and for them [the students] of course, and I am still trying. I struggle to figure out ... you know you are always going to have diverse learning needs in your classroom but adding this language barrier is making it [teaching science] much more difficult because it is very hard to plan a lesson for that many different levels [of English competency]. Because some people are understanding everything you are saying, and other people are understanding 15–20 percent. So, it becomes very ... it's really hard for me to diversify [the instruction] that much because I don't know their [students' English competency] levels very well. It took me a while to figure out the level where people [the students] are at ... [I did] some oral reading [with individual students] to sort of start understanding how much actually they understand. So, it is actually a very difficult thing as a teacher because I am not trained in [teaching] ELLs, right? (Teacher 3)

Second, the teachers were concerned about streaming students into specific science courses, whether through formal placement or student course selection. Formal ESL placements and decisions about ELLs' transition to regular classrooms were made by predominantly White L1 English speakers who seemed to hold deficit-based perceptions about ELLs. The teacher who had herself been a former ESL student conveyed this concern clearly:

> When we have our ESL department meetings, some teachers are going to say that [the student] is too weak [in English] and talk about his [*sic*] being ELL and I am like well, he [*sic*] is weak [in English] but he is really strong academically. So, there is a bit of clash there where I feel like if you can do the work then you should move out [of ESL Science] even if your language [English] is still struggling. (Teacher 3)

Although ESL Science was meant to support ELLs' learning of science while providing additional support in learning English, teachers felt that such placements could negatively impact these students' self-esteem. Teacher 3 had personal experience:

> So, my first language is [non-English]. So, I know that feeling [of being an ESL student] at least early on. I didn't have to spend high school through it [the ESL placement], but it is a struggle, you know. I come from an ESL background as well and I think about these students … that [ESL placement] is really being damaging [for the ELLs]. I think that's a hard thing for people here [in Canada] to see … which is, I mean we have students…. It's really hard for many of us [teachers] to gauge that all students come in here with their own world experiences, their own languages, and they have tremendous amount of knowledge. The only difference between them and our regular students is that they can't find an avenue to express that [knowledge] and that makes it insanely frustrating [for them]. That's why I personally think that some of these [ELLs] have to get out [of the ESL placement] and they have to get out even if they struggle in regular science classes for their own self-esteem. (Teacher 3)

Moreover, English competency level played a central role in the ELLs' enrolment and engagement with specific science subjects. After Grade 10, most ELLs tended to take chemistry or physics because these subjects were less language heavy than biology:

> Because language is an issue for some of the students, their English isn't great, … they tend not to take biology as much because there is so much language in it. Once kids are allowed to choose, the kids who don't have really great English tend to take physics and chemistry more than biology because it's tough to do biology if you don't have a strong command of English. (Teacher 1)

To summarize, the teachers reported that English was the primary mode of instruction in their science classroom although it was a completely new language for many of their students. At times, there were participation and behavioural issues that teachers linked to ELLs' low level of language competency. Thus, they often looked for ways to accommodate ELLs' diverse needs, using a variety of teaching resources and modifying instruction and assessment. The teachers raised concerns about inadequate teacher preparation for working with ELLs and noted that system-based challenges could impact ELLs' self-esteem and influence their selection of upper-level science courses.

4.4 Discussion

In this section, I discuss the themes that arose from my analysis as informed by the framework of (trans-multi)culturally responsive education (Fig. 4.1). The four science teachers reported that the number of ELLs had been increasing at their school, making teaching science more complex. They also reported that a low English competency level created a cultural barrier that negatively affected student socialization and engagement in the science classroom. Research suggests that many ELLs find academic English daunting and that they struggle with everyday English (Cummins et al., 2012), which may explain why ELLs form cultural bubbles that allow them to communicate with peers who spoke the same language. Thus, the teachers enforced the use of English as the only mode of instruction and communication in their science classroom. Their efforts to encourage ELLs to communicate in English are in alignment with findings that students' motivation to learn a new language is influenced by the need to use L2 to communicate (Hollenbeck & Kountouzi, 2019). However, these science teachers' discouragement of L1 use in their classroom is inconsistent with researchers in multicultural science education and culturally responsive literacy practices who advocate for the use of L1 to support ELLs' learning (Cummins et al., 2012; Delpit, 2006; Krugly-Smolska, 2013). Giving students' opportunities to communicate their ideas in L1 and then translate these ideas into academic English enables the development of "code-switching" skills (Ladson-Billings, 1995, p. 161). Although it is difficult to provide bilingual educational opportunities for all student groups, in a multilingual context like the urban Canadian classrooms in this study, teachers could allow ELLs to use their L1 as a cognitive tool (Cummins et al., 2012), and thereby, embrace culturally congruent instruction (Gay, 2010).

Culturally responsive language pedagogy aligns with a (trans-multi)culturally responsive education framework that invites teachers to promote cultural caring and create opportunities for students to build a learning community (Gay, 2010). Moving beyond equality, which assumes that treating everyone the same is enough, a (trans-multi)culturally responsive education advocates for equity (Banks, 2009; Banks & Banks, 2010) to ensure that all students are provided with opportunity to engage in learning and utilize their strengths, such as L1 in the case of ELLs.

Teachers felt that because of low English language competency and prior experiences of being in teacher-centric classrooms ELLs did not actively participate in class discussions and often struggled to engage in inquiry-based science. This finding is consistent with research that suggests that students' cultural norms of respecting teachers as "authoritative sources of knowledge" (Lee, 2003, p. 466) in their home country may discourage them from asking questions and engaging in inquiry-based investigations. Teachers' awareness of students' prior experiences and their willingness to learn about differences in educational systems resonates with CRT that encourages teachers to build their cultural knowledge base (Gay, 2010).

Their efforts to teach inquiry-based science indicate that these teachers wanted ELLs to build these necessary skills; however, they seemed to be prioritizing inquiry as a Eurocentric practice. It is critical to recognize that many cultures have their own rational and empirical ways of knowing nature (Aikenhead & Elliott, 2010). Hence, an inquiry-based approach without culturally congruent instruction may not be sufficient to support ELLs' learning of science (Meyer & Crawford, 2011). A (trans-multi)culturally responsive approach would encourage teachers to engage with ELLs in transcultural dialogues (Keating, 2007) that would help these students in bridging their prior experiences of science and the expectations of Canadian school science. Recognizing and utilizing the linguistic and cultural resources that ELLs bring with them—and explicitly teaching students the cultural codes of power (Barton & Yang, 2000; Delpit, 2006) that are critical for ensuring their successful participation—could help teachers make science inquiry accessible and meaningful for these students (Lee, 2003).

Supporting ELLs' learning in science classrooms can be further complicated by behavioural issues such as disregarding instructions, resisting engagement with peers, and arriving late for class. The teachers reasoned that these behaviours were linked with low English competency. However, the teachers' adherence to English-only instruction disregarded the linguistic capital of ELLs and may have led them to feel unappreciated and disengage from learning (Nieto, 2000), which teachers perceived as disciplinary issues. Teachers' emphasis on English contrasts with a (trans-multi)culturally responsive education perspective that would encourage them to draw on the linguistic capital of students from diverse cultural backgrounds (Nieto, 2000). As such, teachers are called to cultivate critical cultural consciousness to examine their own beliefs (Gay, 2003) and to identify conscious or unconscious biases that are embedded in everyday practices of schooling and larger socio-cultural-political contexts (Ladson-Billings, 2009, 2021) and take actions that may help in reducing prejudice (Banks, 2009; Banks & Banks, 2010). Teachers could interrogate the core causes of student behaviour that may arise from other factors and create a sense of dislocation and isolation among these ELLs.

Considering the varying levels of English language competency, these teachers' efforts to alter their teaching and assessment practices to facilitate ELLs' learning and engagement in their science classroom indicate their willingness to adopt culturally responsive pedagogies. However, some of the strategies they used to support ELLs conflicted with a (trans-multi)culturally responsive education framework. Using decades-old science textbooks to minimize written information and avoiding writing-heavy assignments disregard the fact that learning science goes beyond merely learning facts and vocabulary. These practices in fact limit the opportunities to promote disciplinary literacy among ELLs. As Hynd-Shanahan (2013) noted, disciplinary literacy not only includes content knowledge, but also involves knowledge about how the discipline specific knowledge is generated and communicated within the discipline. Moreover, continued use of these resources is in opposition to nonbinary forms of knowledge, nonbinary epistemological understandings (Keating, 2007), and equity pedagogy (Banks, 2008; Banks & Banks, 2010) that a (trans-multi)culturally responsive education calls teachers to embrace.

System-based challenges also play a role in the complexity of teaching science to ELLs. These participating teachers considered their teacher education program and subsequent professional development inadequate in preparing them to teach

ELLs. All four teachers reported not taking any specific course that focused on supporting ELLs in science contexts, which reflects the reality of many teacher education programs where courses on supporting ELLs are separate from methods courses in core subjects (Lyon et al., 2018). Moreover, secondary teachers generally perceive themselves as content specialists rather than as teachers of language and supporting ELLs at the secondary level is more challenging than at the elementary level because concepts are more abstract (Lyon et al., 2018). In addition, the teachers recognized that the system-based inequities embedded in ESL programs might do more harm than help because they can negatively influence ELLs' self-esteem, hinder their full participation in science classrooms, and impact enrolment in specific science courses. Teachers' concerns about streaming reflect research that suggests that ESL placements often stigmatize nonnative English speakers because their learning abilities are misjudged by their teachers (Lew & Nelson, 2016).

Misjudgment of ELLs' academic potential was raised by these participating teachers. Placement in ESL Science was seen as discriminatory because, despite being capable of navigating regular science courses, many ELLs were automatically placed in ESL Science. Concerns about potential bias in ELLs' placement align with the literature that suggests that many well-meaning White teachers see their culturally diverse students as problems (Moodley, 1995) that need to be fixed (Ladson-Billings, 2001). Looking through a (trans-multi)culturally responsive education perspective, a critical examination of teacher education and professional development programs in developing cultural competency among science teachers (Oliver, 2017) and an ongoing evaluation of the efficacy of ESL programs to support ELLs is essential (Nieto, 2000).

4.5 Conclusion and Implications

In this study, I investigated four secondary science teachers' perspectives about teaching science to ELLs. My findings highlight the complexity of factors that further marginalize already disenfranchised ELLs in Canadian classrooms and call for a closer examination of multicultural education and ESL programs in Canadian schools. My findings have implications for theory, practice, and future research in multicultural science education in Canadian and wider international contexts. The study contributes to the knowledge base in three ways: First, it adds to the literature on complexities in supporting ELLs in secondary science classrooms; second, it proposes a (trans-multi)culturally responsive education framework that can serve as a lens for analyzing teachers' science teaching in diverse Canadian contexts; and, third, it brings Canadian teachers' voices to the fore.

Because teachers' domain-specific science knowledge is insufficient for promoting scientific literacy among diverse students including ELLs (Brown & Aydeniz, 2017), it is critical that science teachers have an understanding of disciplinary literacy so as to identify and model language features that could aid students to learn science content (Turkan et al., 2014). Teachers need to actively work toward cultivating disciplinary literacy that varies according to the ways of thinking and processes of knowledge generation that are characteristic of the discipline (Hynd-Shanahan, 2013). Hence, rather than eliminating the language challenges that are involved in learning science (Lyon et al., 2018), teachers should employ language-intensive science literacy practices that could support the learning of both science and language (Lee et al., 2013). Teacher education programs and professional development should support teachers in acquiring a cultural knowledge base and understandings that facilitate effective instruction to support ELLs' learning in their schools (Cummins et al., 2012).

The (trans-multi)culturally responsive education framework encourages teachers to cultivate relational connectedness and engage with students in transcultural dialogues (Keating, 2007) to co-construct science inquiries that acknowledge the contributions of all cultures and associated languages. By encouraging the use of linguistic capital that ELLs bring with them, (trans-multi)culturally responsive science teachers embrace diversity and difference as valid vehicles for learning (Nieto, 2000). They engage in ongoing critical self-reflections to identify how the issues of prejudice, discrimination, and marginalization might be systemically infused in their everyday pedagogical practices, classrooms, and school cultures (Barton & Yang, 2000).

To establish an empowering school culture (Banks, 2009; Banks & Banks, 2010), a (trans-multi)culturally responsive education encourages (w)holistic transformation of educational processes where the entire culture of classrooms and schools is critically examined to inquire:

- How is linguistic diversity implicated in everyday science instruction?
- What inequities exist in ESL placement policies?
- How does the lack of adequate teacher preparation and professional development contribute to challenges of supporting ELLs in science classrooms?

Acknowledgements I dedicate this chapter to many ELLs who continue to long for a sense of belongingness and dignified integration of their linguistic diversity in contemporary science classrooms. I thank all the participating teachers without whose contributions this research would not have been possible. I express my sincere gratitude to Dr. Samson M. Nashon, Dr. Hartej Gill, and Dr. Stephen Petrina for their invaluable guidance, support, and encouragement provided throughout this research.

References

Aikenhead, G. S., & Elliott, D. (2010). An emerging decolonizing science education in Canada. *Canadian Journal of Science, Mathematics & Technology Education, 10*(4), 321–338. https://doi.org/10.1080/14926156.2010.524967

Arkoudis, S. (2003). Teaching English as a second language in science classes: Incommensurate epistemologies. *Language & Education, 17*(3), 161–173. https://doi.org/10.1080/09500780308666846

Banks, J. A. (2009). Multicultural education: Dimensions and paradigms. In J. A. Banks (Ed.), *Routledge international companion to multicultural education* (pp. 9–32). Routledge. https://doi.org/10.4324/9780203881514

Banks, J. A., & Banks, C. A. M. (2010). *Multicultural education: Issues and perspectives* (10th ed.). Wiley.

Barton, A. C., & Yang, K. (2000). The culture of power and science education: Learning from Miguel. *Journal of Research in Science Teaching, 37*(8), 871–889. https://doi.org/10.1002/1098-2736(200010)37:8%3c871::AID-TEA7%3e3.0.CO;2-9

Brown, C. L., & Aydeniz, M. (2017). Maximizing science teachers' pedagogical potential for teaching science literacy to ELLs: Findings from a professional development program. In L. C. de Oliveira & K. Campbell Wilcox (Eds.), *Teaching science to English Language Learners: Preparing pre-service and in-service teachers* (pp. 83–100). Springer. https://doi.org/10.1007/978-3-319-53594-4_5

Callaham, R. M., Sampson, V., & Rivale, S. (2019). Activating bilingual English Language Learners' strengths in science: The pedagogy of argument driven inquiry (ADI). In L. C. de Oliveira, K. M. Obenchain, R. H. Kenney, & A. W. Oliveira (Eds.), *Teaching the content areas to English Language Learners in secondary schools: English language arts, mathematics, science, and social studies* (pp. 183–197). Springer. https://doi.org/10.1007/978-3-030-02245-7_12

Cardozo-Gaibisso, L., Allexsaht-Snider, M., & Buxton, C. A. (2017). Curriculum in motion for English Language Learners in science: Teachers supporting newcomer unaccompanied youth. In L. C. de Oliveira & K. Campbell Wilcox (Eds.), *Teaching science to English Language Learners: Preparing pre-service and in-service teachers* (pp. 7–29). Springer. https://doi.org/10.1007/978-3-319-53594-4_2

Chui, T. W. L., & Flanders, J. (2013). *Immigration and ethnocultural diversity in Canada: National household survey, 2011* (Catalogue no. 99–010-X2011001). Statistics Canada. https://www12.statcan.gc.ca/nhs-enm/2011/as-sa/99-010-x/99-010-x2011001-eng.pdf

Council of Ministers of Education, Canada. (1997). *Common framework of science learning outcomes, K to 12: Pan-Canadian protocol for collaboration on school curriculum for use by curriculum developers*. https://archive.org/stream/commonframework00coun/1996330#mode/2up

Cummins, J., Mirza, R., & Stille, S. (2012). English Language Learners in Canadian schools: Emerging directions for school-based policies. *TESL Canada Journal, 29*(Special issue 6), 25–48. https://doi.org/10.18806/tesl.v29i0.1121

Delpit, L. D. (2006). *Other people's children: Cultural conflict in the classroom*. New Press.

Egbo, B. (2019). *Teaching for diversity in Canadian schools* (2nd ed.). Pearson Canada Inc.

Gay, G. (Ed.). (2003). *Becoming multicultural educators: Personal journey toward professional agency*. Jossey-Bass.

Gay, G. (2010). *Culturally responsive teaching: Theory, research, and practice* (2nd ed.). Teachers College Press.

Ghosh, R., & Galczynski, M. (2014). *Redefining multicultural education: Inclusion and the right to be different* (3rd ed.). Canadian Scholars' Press.

Gill, H., & Chalmers, G. (2007). Documenting diversity: An early portrait of a collaborative teacher education initiative. *International Journal of Inclusive Education, 11*(5–6), 551–570. https://doi.org/10.1080/13603110600600986

Goodenough, W. H. (1976). Multiculturalism as the normal human experience. *Council on Anthropology & Education Quarterly, 7*(4), 4–7. https://doi.org/10.1525/aeq.1976.7.4.05x1652n

Henry, A. (2017). Culturally relevant pedagogy in Canada: Possibilities and challenges for African Canadian children. *Teachers College Record, 119*(1), 1–27. https://doi.org/10.1177/016146811711900103

Hollenbeck, J. E., & Kountouzi, B. (2019). Using motivation in the instruction of English language learning for STEM and economic secondary education for Bulgarian learners. *Bulgarian Journal of Science & Education Policy, 13*(1), 71–79. Retrieved from https://www.proquest.com/openview/11e6cc55958a6efdfd77a2b81e66fdaf/1?pq-origsite=gscholar&cbl=1056402

Horowitz, F. D., Darling-Hammond, L., Bransford, J., Comer, J., Rosebrock, K., Austin, K., & Rust, F. (2005). Educating teachers for developmentally appropriate practice. In L. Darling-Hammond & J. Bransford (Eds.), *Preparing teachers for a changing world: What teachers should learn and be able to do* (pp. 88–125). Jossey-Bass.

Hynd-Shanahan, C. (2013). What does it take? The challenge of disciplinary literacy. *Journal of Adolescent & Adult Literacy, 57*(2), 93–98. https://doi.org/10.1002/JAAL.226

Joshee, R., Peck, C., Thompson, L. A., Chareka, O., & Sears, A. (2016). Multicultural education, diversity, and citizenship in Canada. In J. Lo Bianco & A. Bal (Eds.), *Learning from difference: Comparative accounts of multicultural education* (pp. 35–50). Springer. https://doi.org/10.1007/978-3-319-26880-4_3

Keating, A. (2007). Teaching transformation: Transcultural classroom dialogues. *Palgrave Macmillan*. https://doi.org/10.1057/9780230604988

Kelly, G. J., Brown, B., & Jiménez-Aleixandre, M. P. (2023). Discourse practices in science learning. In N. G. Lederman, D. L. Zeidler, & J. S. Lederman (Eds.), *Handbook of research on science education* (Vol. III, pp. 413–446). Routledge. https://doi.org/10.4324/9780367855758-18

Krugly-Smolska, E. (2013). Twenty-five years of multicultural science education: Looking backward, looking forward. *Encounters on Education, 14*(Fall), 21–31. https://doi.org/10.24908/eoe-ese-rse.v14i0.4522

Ladson-Billings, G. (1995). But that's just good teaching! The case for culturally relevant pedagogy. *Theory into Practice, 34*(3), 159–165. https://doi.org/10.1080/00405849509543675

Ladson-Billings, G. (2001). *Crossing over to Canaan: The journey of new teachers in diverse classrooms*. Jossey-Bass.

Ladson-Billings, G. (2009). *The dreamkeepers: Successful teachers of African American children*. Jossey-Bass Publishers.

Ladson-Billings, G. (2021). *Culturally relevant pedagogy: Asking a different question*. Teachers College Press.

Lee, O. (2003). Equity for culturally and linguistically diverse students in science education: A research agenda. *Teachers College Record, 105*(3), 465–489. https://doi.org/10.1111/1467-9620.00247

Lee, O. (2005). Science education with English language learners: Synthesis and research agenda. *Review of Educational Research, 75*(4), 491–530. https://doi.org/10.3102/00346543075004491

Lee, O., Quinn, H., & Valdés, G. (2013). Science and language for English language learners in relation to next generation science standards and with implications for common core state standards for English language arts and mathematics. *Educational Researcher, 42*(4), 223–233. https://doi.org/10.3102/0013189X13480524

Lee, S. (2010). Reflect on your history: An early childhood education teacher examines her biases. *Multicultural Education, 17*(4), 25–30. https://files.eric.ed.gov/fulltext/EJ915268.pdf

Lew, M. M., & Nelson, R. F. (2016). New teachers' challenges: How culturally responsive teaching, classroom management, & assessment literacy are intertwined. *Multicultural Education, 23*(3–4), 7–13. https://files.eric.ed.gov/fulltext/EJ1119450.pdf

Lyon, E. G., Stoddart, T., Bunch, G. C., Tolbert, S., Salinas, I., & Solis, J. (2018). Improving the preparation of novice secondary science teachers for English learners: A proof of concept study. *Science Education, 102*(6), 1288–1318. https://doi.org/10.1002/sce.21473

Marton, F. (1986). Phenomenography—A research approach to investigating different understandings of reality. *Journal of Thought, 21*(3), 28–49. https://ia803400.us.archive.org/27/items/sim_journal-of-thought_fall-1986_21_3/sim_journal-of-thought_fall-1986_21_3.pdf

Marton, F. (2014). Necessary conditions of learning. *Routledge*. https://doi.org/10.4324/9781315816876

Merriam, S. B. (1998). *Qualitative research and case study applications in education* (2nd ed.). Jossey-Bass.

Meyer, X., & Crawford, B. A. (2011). Teaching science as a cultural way of knowing: Merging authentic inquiry, nature of science, and multicultural strategies. *Cultural Studies of Science Education, 6*(3), 525–547. https://doi.org/10.1007/s11422-011-9318-6
Miles, M. B., & Huberman, A. M. (1994). *Qualitative data analysis: An expanded sourcebook* (2nd ed.). Sage.
Milford, T. M., & Tippett, C. D. (2016). Canada: Teaching science in Canada. In B. Vlaardingerbroek & N. Taylor (Eds.), *Teacher quality in upper secondary science education: International perspectives* (pp. 43–58). Palgrave Macmillan.
Moodley, K. A. (1995). Multicultural education in Canada: Historical development and current status. In J. A. Banks & C. A. M. Banks (Eds.), *Handbook of research on multicultural education* (pp. 801–820). Simon & Shuster Macmillan.
Nashon, S. M., & Anderson, D. (2013). Interpreting student views of learning experiences in a contextualized science discourse in Kenya. *Journal of Research in Science Teaching, 50*(4), 381–407. https://doi.org/10.1002/tea.21078
Nieto, S. (2000). *Affirming diversity: The sociopolitical context of multicultural education* (3rd ed.). Longman.
Oliver, G. (2017). Supporting English Language Learners in secondary science through culturally responsive teaching. In L. C. de Oliveira & K. Campbell Wilcox (Eds.), *Teaching science to English Language Learners: Preparing pre-service and in-service teachers* (pp. 111–136). Springer. https://doi.org/10.1007/978-3-319-53594-4_6
Olson, J. K., Tippett, C. D., Milford, T. M., Ohana, C., & Clough, M. P. (2015). Science teacher preparation in a North American context. *Journal of Science Teacher Education, 26*(1), 7–28. https://doi.org/10.1007/s10972-014-9417-9
Raisinghani, L. (2018). *Teachers' perspectives on culturally diverse classrooms and responsive science and mathematics teaching* [Doctoral dissertation, University of British Columbia]. https://doi.org/10.14288/1.0372169
Raisinghani, L. (2019, September 18). (Trans-multi)culturally responsive education: A critical framework for responding to student diversity. *Education Canada: Cutting through the rhetoric, 59*(3), 26–31. https://www.edcan.ca/articles/trans-multiculturally-responsive-education/
Reid, J. W., Smith-Walters, C., Mangione, K. A., Dorris, A., & Tharp, T. (2019). Supporting English language learners through inquiry-based science: Three strategies for your classroom. In L. C. de Oliveira, K. M. Obenchain, R. H. Kenney, & A. W. Oliveira (Eds.), *Teaching the content areas to English Language Learners in secondary schools: English language arts, mathematics, science, and social studies* (pp. 199–214). Springer. https://doi.org/10.1007/978-3-030-02245-7_13
Stake, R. E. (2000). Case studies. In N. K. Denzin & Y. S. Lincoln (Eds.), *Handbook of qualitative research* (2nd ed., pp. 435–454). Sage.
Statistics Canada. (2022, October 26). *The Canadian census: A rich portrait of the country's religious and ethnocultural diversity* [Press release]. https://www150.statcan.gc.ca/n1/en/daily-quotidien/221026/dq221026b-eng.pdf?st=VWuccxiC
Teel, K. M., & Obidah, J. E. (Eds.). (2008). *Building racial and cultural competence in the classroom: Strategies for urban educators*. Teachers College Press.
Thomas, E. (2003). The case for a culture sensitive education, and the role of pedagogy in resolving tensions between tradition, modernity and postmodernity. In M. W. Charney, B. S. A. Yeoh, & T. C. Kiong, (Eds.), *Asian migrants and education: The tensions of education in immigrant societies and among migrant groups* (pp. 65–81). Springer. https://doi.org/10.1007/978-94-017-0117-4_6
Turkan, S., de Oliveira, L. C., Lee, O., & Phelps, G. (2014). Proposing a knowledge base for teaching academic content to English language learners: Disciplinary linguistic knowledge. *Teachers College Record, 116*(3), 1–30. https://doi.org/10.1177/016146811411600303
Vancouver School Board. (2024). *Our district*. https://www.vsb.bc.ca/page/4775/our-district

Vézina, M., & Maheux, H. (2017, October 25). *Ethnic and cultural origins of Canadians: Portrait of a rich heritage. Census of population, 2016* (Catalogue no. 98–200-X2016016). Statistics Canada. https://www12.statcan.gc.ca/census-recensement/2016/as-sa/98-200-x/2016016/98-200-x2016016-eng.pdf

Chapter 5
Changes in Science Teacher PCK During Grade 10 Ocean Inquiry Activities

Patrick R. Wells and Joy Ricketts

5.1 Introduction

In Canada, the ocean infuses the arts, history, and countless industries (Glithero, 2020). However, few people appreciate how the ocean makes the Earth habitable (Fauville et al., 2018; Strang et al., 2007). Tran et al. (2010) suggested the development of an ocean-literate society requires that "ocean and aquatic sciences must be valued and integrated into educational practice, research, standards, curriculum, textbooks, and assessment" (p. 22). The ocean can be an engaging context for science instruction, and many science topics arguably cannot be understood without knowledge of the ocean (Strang et al., 2007). For example, any discussion of climate, including climate change, must include the ocean. Indigenous knowledge holders, ocean educators, and marine scientists understand that the ocean's abiotic and biotic environments play an important role in sustainability through the connection to weather and climate (Han et al., 2019). Specifically, in Canada's eastern coastal province of Newfoundland and Labrador, sustainability is a concern for the recovering northern cod stocks (Rose, 2018), marine plastic pollution (Liboiron, 2021), and changes in Labrador coastal ice amount and quality (Bell et al., 2014). Climate change and sustainability highlight the importance of engaging a broader audience, such as Kindergarten to Grade 12 students, with marine science and ocean learning (Worm et al., 2021).

Ocean lessons can provide a powerful science learning context while increasing ocean literacy (Guest et al., 2015; Putland et al., 2021). Studies in British Columbia

P. R. Wells (✉)
Memorial University of Newfoundland, St. John's, NL, Canada
e-mail: p.wells@mun.ca

J. Ricketts
Roncalli Central High, Avondale, NL, Canada
e-mail: joyricketts@nlschools.ca

C. D. Tippett and T. M. Milford (eds.), *Seeing Science Through the Eyes of Canadian Teachers and Learners*, Contemporary Trends and Issues in Science Education 54,
https://doi.org/10.1007/978-3-031-88785-7_5

(Putland et al., 2021) and Newfoundland and Labrador (Wells & Negirjn, 2019) have demonstrated that the ocean can be an influential context for Grades 8–12 students' science learning. Research also suggests that ocean-literate citizens can link human activities to rising sea levels and envision the ecological and economic importance of protected marine areas (Fauville et al., 2018; Guest et al., 2015). Monroe et al. (2019) suggested that effective climate change programs can be "personally relevant and meaningful for learners … [with] activities or educational interventions … designed to engage learners." (p. 799). Teachers can support students' ocean literacy development through an inquiry approach to topics such as ocean currents (Reckinger et al., 2014), ocean acidification (Erickson & Crews, 2016; Kelley et al., 2015), and the impact of climate change (National Oceanic & Atmospheric Administration [NOAA], 2021).

5.1.1 The Scarcity of High School Ocean Literacy Resources

Unfortunately, curricula and textbooks with a terrestrial bias (Fauville, 2018; Stewart, 2021) limit the contexts for ocean literacy for Canadian students. Many students in Newfoundland and Labrador live close to the ocean and are regularly influenced by abiotic and biotic components (Banfield, 1993). Thus, we decided to develop ocean activities that could engage and empower our students while expanding their knowledge of the ocean's impact on them and their impact on the ocean. These inquiry activities would be place-focused and strive to be inclusive (Arenas, 2001; Smith, 2002; Sobel, 2004). An inquiry approach would help students meet the seminal goals of science instruction: "learning science, learning about science, doing science, and addressing socio-scientific issues" (Hodson, 2014, p. 2537). During inquiry-based learning, students use skills and practices that "scientists employ as they investigate and build models and theories about the world" (NRC, 2012, p. 30). Conducting inquiry can motivate learners and has been linked to improved performance on high-stakes tests (Saunders-Stewart et al., 2012). Compared to didactic instruction, inquiry is more effective and engaging for students (Wilson et al., 2010) thereby making inquiry a more inclusive pedagogy (Adu-Boateng & Goodnough, 2022; Florian & Beaton, 2018). Our ocean activities would be hands-on, with multisensory observation opportunities and less text, to include and support more students (Carp, 2007; Clarke & Kast, 2020). Placed-based learning can be engaging (Sobel, 2004); when framed appropriately, local science inquiry can foster global connections, so we wanted to draw on the marine setting of Newfoundland and Labrador (Arenas, 2001).

5.1.2 Science Teacher Pedagogical Content Knowledge

Pedagogical content knowledge (PCK) is the theoretical framework for teacher learning of this study and was originally conceptualized as "an understanding of how particular topics, problems, or issues are organized, represented, and adapted to the diverse interests and abilities of learners, and presented for instruction" (Shulman, 1987, p. 8). The Consensus Model of PCK in Science Education (Gess-Newsome, 2015) was recently followed by the Refined Consensus Model of PCK in Science Education (RCM; Carlson et al., 2019; Fig. 5.1). The RCM includes five forms of teacher knowledge and skill components: content knowledge, pedagogical knowledge, knowledge of students, curricular knowledge, and assessment knowledge. During the COVID-19 pandemic, Wells et al. (2023) used the RCM to document changes in distance education science teachers' PCK during the development of remote lessons.

Within the learning context, there are two forms of PCK: personal PCK (pPCK) and enacted PCK (ePCK). The totality of a science teacher's knowledge is their pPCK, the "cumulative and dynamic pedagogical content and skills for teaching" that includes teacher and student contributions (Carlson et al., 2019, p. 85). During the ocean lessons in our investigation, ePCK was "the specific knowledge and skills utilized by an individual teacher" (p. 83) with the students in our classroom. A teacher's ePCK is engaged during enactment when pedagogical reasoning is employed for teaching, planning, and reflecting (Fig. 5.1). Furthermore, within pedagogical reasoning, there are two forms of reflection: reflection *in action,* which occurs with students during instruction, and reflection *on action,* which recalls prior instructional events such as previous student interactions during lesson planning

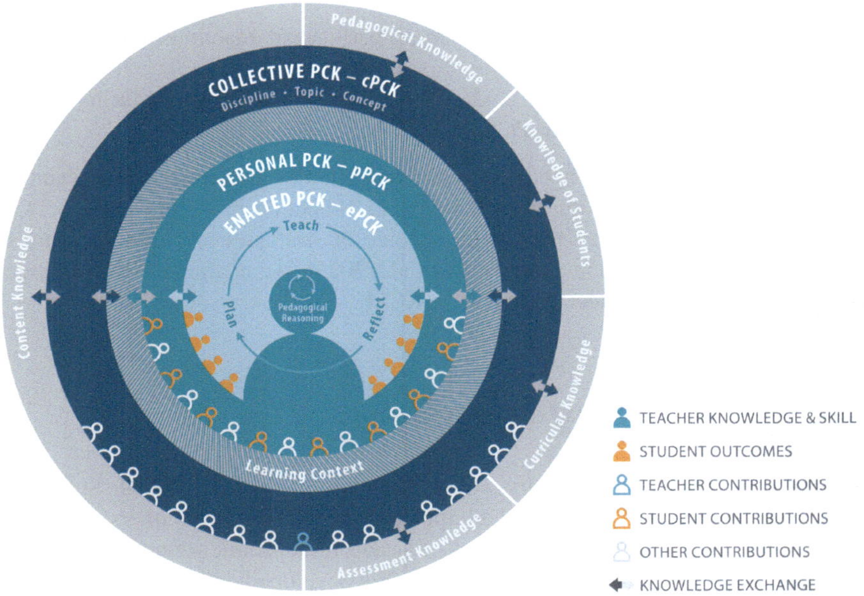

Fig. 5.1 The refined consensus model of pedagogical content knowledge in science education (Carlson et al., 2019)

(Carlson et al., 2019). Collective PCK (cPCK) is the discipline, topic, and concept knowledge of all science teachers. Each form of PCK exchanges knowledge with the five components, or knowledge bases, of knowledge and skill for science teaching (Fig. 5.1).

5.2 Method: An Action Research Approach

To explore aspects of PCK, we took an action research approach, which is a reflective form of practitioner research (Dana & Yendol-Hoppey, 2014) that has fostered the development of a science inquiry community of practice in our school (Wells et al., 2017) and more broadly in the province (Goodnough et al., 2010). We collaboratively examined our classroom practice to generate knowledge using a process that "engages teachers in the design, data collection, and interpretation of data around a question" (Dana & Yendol-Hoppey, 2014, p. 8). Our action research learning cycle included lesson development, teaching data collection, and data analysis that parallels the plan–teach–reflect contexts of the RCM with reflection in and on action (Carlson et al., 2019).

We initially considered how locally focused ocean inquiry activities might impact students' understanding of the ocean's influence on weather and climate. However, we refined this query to develop more specific questions to guide the action research cycles:

1. What teacher PCK is developed when preparing and enacting inquiry-focused ocean education lessons?
2. How did our perceptions of student engagement and learning during ocean-inquiry activities impact our PCK?

During our action research, our collective reflection on action for lesson planning and pedagogical decisions during teaching (i.e., reflection in action) makes ePCK visible as an "expression of knowledge, choice of strategies and representations" (Carlson et al., 2019, p. 85). Like Wells et al. (2023), we examined PCK changes within the plan–teach–reflect stages of the RCM.

5.2.1 Research Context and Participants

This study took place in a secondary school located in an urban area outside St. John's, the provincial capital. The school had approximately 50 staff members and about 750 students. We co-taught two sections of Grade 10 academic science course with a Sustainability and Weather Dynamics Unit. We developed and implemented three hands-on ocean inquiry activities that addressed curriculum outcomes and linked with Ocean Literacy Principle (OLP) 3: "The ocean is a major influence on weather and climate" and OLP 6: "The ocean and humans are inextricably interconnected"

(Santoro et al., 2017, p. 19). We considered student manipulation of local marine samples, including shells and iceberg ice, to be mutually important for engagement and memory formation (Kelly et al., 2022; Monroe et al., 2019). All student inquiries used Vernier® interfaces and probes to quantify and graph simultaneously student-sensed phenomena. This use of technology provided students "more time to observe, to reflect, and to construct conceptual knowledge that underlies the laboratory experiences" (Hofstein & Lunneta, 2004, p. 41).

5.2.1.1 Teachers

The teacher participants were the authors of this chapter. The first author, Pat, was a science subject area specialist with a BSc, a master's in biology, a BEd, and had 24 years of science instruction experience. Pat was a member of several provincial curriculum writing committees and was the science content and curriculum lead for the two science classes of this study. The second author, Joy, was the instructional resource teacher for both classes and supported student learning. Joy had advanced training in the Department of Education Service Delivery Model, a BSpEd, a master's in folklore/special education, and was the lead for student cognitive, behavioural, and social support. Over the six months of shared instruction, we developed a co-teaching relationship that was maintained and negotiated during frequent after-class meetings to review lessons and clarify roles and responsibilities.

5.2.1.2 Students and Recruitment

Joy noted that our classes "had a lot of diverse learners." This statement illustrates her knowledge of the individual education plans (IEPs) for 14 students in the two classes of 28 and 30 students. Recruitment for this study was anonymous and conducted by a third party who, in our absence, reviewed school district ethical protocols regarding the students' and parents' right to refuse participation. All students had access to the district-approved informed consent forms, and the third party collected forms in a secure drop box that was not opened until the final grades were submitted. At year-end, 18 students consented to participate in the study, with 14 parents/guardians consenting to the examination laboratory reports and using direct quotes from those documents. The participants were 15 and 16 years old with final marks ranging from 62 to 94%. Three students in the sample had IEPs, suggesting that the participating students had diverse levels of sensory and/or cognitive ability.

5.2.2 The Ocean Inquiry Activities

For the three ocean inquiry interventions, although students were supplied with materials and procedures, they were required to make predictions, collect data, and present data to support or refute their predictions.

1. *Investigating Albedo*. Students predict and then determine the change in temperature of differently coloured fabrics exposed to radiant energy.
2. *Studies of Sea Water and Iceberg Water*. Students determine the quality of iceberg water (i.e., glacial ice) and local seawater then conduct ocean acidification experiments.
3. *Modelling the Ocean with Glacial Ice*. Students model the impact of iceberg ice and radiant energy on ocean currents.

The students worked in laboratory groups with roles that rotated for each inquiry: Leader (1 student), Equipment Manager (2 or more students), and Data Manager (2 or more students). The rotation of roles ensured that each student experienced the duties, skills, and responsibilities of each role. All groups were heterogeneous regarding student academic and sensory abilities, and the activities included opportunities for multisensory engagement and interactions where possible.

The lessons were designed to provide opportunities to examine planetary-scale phenomena, learn about the ocean's impact on weather and climate (OLP 3), and understand how humans are connected to the ocean (OLP 6). The foci of the action research for ocean activities included place, student engagement, and process skills for science inquiry. The activities and the laboratory group structure fostered elements of a feedback culture (Hattie & Clarke, 2018), and each activity included a task-related formative assessment framework of predictions within the mixed-ability groups that fostered equity.

5.2.3 Data Collection and Analysis

We used several forms of data to strengthen data source and type triangulation (Creswell, 2015). Data included teacher lesson plans, notes and observations of the enactment of each activity, and post-lesson debriefs to adjust the activities for future use, including written reflections that suggested lesson modifications. During instruction, we observed each other and the students, then met to reflect on each activity and its impact based on student engagement, their in-class questions or comments, and learning based on their written laboratory reports. Other student data included fieldnotes collected during lessons that were limited to annotations of key events with students and were written after the class was completed. We also examined the written laboratory reports of the 14 consenting students.

The data were uploaded to MAXQDA (https://www.maxqda.com/) and coded with a priori codes from the RCM (Carlson et al., 2019), the stages of action research

(Dana & Yendol-Hoppey, 2014), and components of OLPs 3 and 6. However, we found that the pedagogical reasoning stages (i.e., plan–teach–reflect) were most useful for examining each activity during development, enactment, and reflection.

5.3 Results

According to Carlson et al. (2019), the contexts of Plan and Reflect engage reflection on action, while the Teach phase sometimes requires reflection in action. In the Plan phase, we found our intended lesson actions activated existing ePCK and developed new ePCK for the enactment of the Teach phase. Teaching provided live feedback and required new reflection on action in response to student outcomes. The pedagogical reasoning of Plan and Teach was examined in Reflect, along with the student laboratory reports. Each contextually unique stage employed pedagogical reasoning; however, the decision to retain new ePCK into pPCK employed reflection on action of the student outcomes and our perceived effectiveness of the lessons (summarized in Table 5.1). The subsequent analysis provides details of our pedagogical reasoning, by stage, for each inquiry lesson.

5.3.1 Plan

Planning the inquiry classes involved meetings after class and school to discuss the lessons and outcomes that would be addressed in the following days. During planning, we engaged our ePCK and drew on the cPCK from publications and websites to support areas of content knowledge and pedagogical strategies not found in the curriculum guide. Each of us brought different skills to the co-teaching team. Pat was involved in developing the course curriculum and taught this version of the course for 16 years, while Joy had three years of working as a science support teacher. As a result, she picked up the curriculum and content knowledge ePCK when preparing for the lessons and was invaluable support for Pat as a student behaviour and cognitive ability expert. During planning, Joy frequently reminded Pat of student capabilities, IEP requirements, and strategies for optimal instruction thereby adding to Pat's ePCK for knowledge of students. Although we had different perspectives, we both agreed that social learning with an ocean focus and data collection inquiry activities would likely engage many students and address our action research goal.

5.3.1.1 Plan—Laboratory Practices and Routines

The students were habituated to using Google Classroom to view inquiry protocols such as safety and group role expectations. We agreed that the activities would include our use of student predictions, an important student engagement and

Table 5.1 Summary of topic-specific ePCK developed across plan–teach–reflect stages of action research

Teacher	Plan Reflection on action	Teach Reflection in action	Reflect Reflection on action
Joy	CK—ocean-related (water pH & density), knowledge from pPCK sources PK—inquiry and technology skills (new discrepant event, temperature probe) CurrK—curriculum outcomes KA—formative assessment and summative report questions	CK—graphical relationships for heating in albedo laboratory PK—application of inquiry skills in new context (new discrepant events, temperature probe, pH probe), decision making during instruction KS—motivation, misconceptions KA—formative assessment with comments, support for summative report questions SO—student engagement with laboratory-discrepant events, observations using multiple senses, data collection	CK—ocean-related knowledge from pPCK sources from student reports PK—new inquiry and technology skills KS—use more local content KA—clarify directions for summative questions SO—student report laboratory misconceptions, importance of discrepant events, sensory observations during data collection
Pat	CK—new ocean knowledge from cPCK sources PK—application of inquiry skills to develop new lessons (new discrepant event) KS—sensory and cognitive needs, learning difficulties KA—formative assessment and summative report questions	CK—heat capacity of seawater, glacial ice properties PK—application of inquiry skills in new context (new discrepant events, pH probe), decision making during instruction KS—motivation, misconceptions, needs KA—formative assessment with comments, support for summative report questions SO—student engagement with laboratory-discrepant events, observations using multiple senses, data collection	CK—ocean-related knowledge from pPCK sources from student reports PK—retain effective aspects of new inquiry and technology skills, modify the ocean model KS—use more local content KA—clarify directions for summative questions SO—student report laboratory misconceptions, importance of discrepant events, sensory observations during data collection

Note AK = assessment knowledge; CK = content knowledge; CurrK = curricular knowledge; KS = knowledge of students; PK = pedagogical knowledge; SO = student outcomes

formative assessment strategy that we found assists in both finding and addressing misconceptions.

The students had experience using technology, collecting data, examining tables and graphs of data, and sharing results through screenshots or Google Docs. The ocean acidification inquiry used the pH probe that Joy was familiar with from previous chemistry activities; during planning for the ocean current lessons, Joy was briefed on the use of the new temperature probe. Pat quickly demonstrated, "So, you plug it in as with all the rest [of the probes] and the LabQuest 2 picks it up. Hit run and collect some data." Pat then squeezed the temperature probe in his hand, and the interface graph demonstrated a rise in temperature. "That's good!" said Joy. Our skills for troubleshooting and managing technology issues were well established [pPCK; Joy's new inquiry skill was ePCK, shared from Pat's pPCK].

5.3.1.2 Plan—Albedo Inquiry

The albedo activity involved temperature changes of different fabrics with varying levels of reflectiveness when exposed to infrared radiation from a heat lamp. Students collected data using both alcohol thermometers and the Vernier® temperature probes. This activity might be considered confirmation inquiry (Banchi & Bell, 2008); however, for students not familiar with albedo or who hold misconceptions, the investigation was not confirmation. Pat conducted this activity with students; in previous iterations, he had found explicit reminders of the definition of albedo supported students when they made predictions connecting albedo and temperature change based on the fabric colour when exposed to infrared radiation (pPCK). We added post-inquiry questions to connect albedo with climate change (Sagan et al., 1979) and weather and the near ocean (Comiso, 2001). These elaboration questions were in the student report:

- Due to the increased levels of carbon dioxide within the atmosphere, the global temperature has been rising for the past 100 years. What impact would the loss of the glaciers have on the temperature of the planet?
- In summer when the wind blows from the northwest across Conception Bay, the dark blue ocean with its low albedo does not significantly warm the air. Suggest two reasons why the air remains cool.

We thought that discussing these questions would help students in future ocean current and melting glacier ice activities [ePCK using new knowledge, summative assessment, reflection on action].

5.3.1.3 Plan—Ocean Acidification Inquiry and Discrepant Event

Pat suggested that the ocean acidification inquiry should start with a CO_2 discrepant event (Hoover, 2017) to demonstrate the impact of acid on locally collected clam shells. We tested 1.0 M HCl and found it sufficient to release CO_2 gas [a new ePCK for chemical reactions]. Before dropping the shell in the acid, we asked the students to predict, "What will happen when this shell is dropped into a clear solution?".

This demonstration would also use the flame test used in previous chemistry inquiries to demonstrate that the evolved gas was CO_2, H_2, or O_2 [pPCK]. Another planned discrepant was pouring the CO_2 produced in the reaction on a candle flame to demonstrate the "fluid" nature of gases; this extinguished the flame during testing [new ePCK for instruction].

Following the discrepant events, the ocean acidification inquiry had two separate investigations: clamshell decomposition in three fluids (acidic, neutral, and basic) and seawater pH values as CO_2 is bubbled through it. We planned to have groups share data the day after the inquiry was completed. Next, to make connections between observations and CO_2 planetary processes, we discussed a popular ocean acidification infographic from the NOAA (2021).

The idea for exposing seashells to acid was Pat's adaptation of a lesson from Gorospe et al. (2013). During lesson planning, Pat introduced Joy to ocean acidification and the connection between increased atmospheric carbon dioxide and the lowering of ocean pH (increasing acidity). Pat's ocean acidification knowledge was drawn from websites and scientific papers accessed while planning: using cPCK sources to develop ePCK content knowledge during planning. Pat's ePCK that ocean water has a pH of 8 or higher also supported a discussion with Joy when tests of local ocean water pH were 7.7, much higher than the pond water studied in the fall [new ePCK content knowledge for Joy].

5.3.1.4 Plan—Ocean Currents and Icebergs

The ocean currents activity was inspired by ocean modelling research on the impact of the Earth's rotation on currents (Reckinger et al., 2014). The inquiry ocean model consisted of a 38L aquarium with 20L of 4 °C seawater, an iceberg made with chunks of iceberg ice, and a radiation source (heat lamp). We wanted two identical ocean models, except that one would have alcohol thermometers and the other digital thermometers connected to a Vernier LabQuest2®. Data collection would start with a 5-min baseline reading of the seawater. Then, students would predict possible temperature changes as the heat source was turned on, and a chunk of iceberg ice measuring −10° was carefully lowered into the water. Temperatures would be documented for 30 min.

This activity was exploratory for both the students and teachers as we did not pretest the apparatus. We know the density of seawater increases with salinity, that cold-denser seawater falls in a water column, and that warming water is less dense and rises; but as with any untested model, we were unsure what would happen.

5.3.2 Teach

5.3.2.1 Teach—Teaching and Reflection in Action

Each activity started in a similar manner with a flurry of activity as groups collected their logbooks and determined their rotating laboratory role. The equipment managers collected the required LabQuest2® sensors, booted up the data collection interface, and selected and set up the equipment for the activity. The data managers secured two iPads, logged one into Google Classroom and connected the other to the private wireless network of the data collection interface. The laboratory leaders checked with the equipment managers to ensure they were ready to conduct the experiment and that the data managers were prepared to share the data and observations with all group members using Google Docs.

5.3.2.2 Teach—Learning During Enactment of the Albedo Inquiry

The albedo laboratory collected data to examine albedo's impact on environmental temperature change (Sagan et al., 1979). Albedo is a measure of reflectiveness; from 10 years of conducting the albedo laboratory, Pat was aware that the main student misconception was confusing absorption and reflectiveness [pPCK]. Students placed temperature probes under fabrics, each with a different albedo. Data collection, along with explicit reminders of the definition of albedo, helped to address the main albedo misconception and questions that supported student learning, "Don't forget, albedo is a measure of reflectiveness, and white has the highest albedo."

Following calibration, the lamp's radiant energy was transferred to the coloured fabrics. The students watched the interface graph production, knowing which line was associated with the highest and lowest albedo (i.e., white and black, respectively). As we moved through the groups, we detected no issues with conducting the experiment and added our explicit reminders and knowledge probes: "What are you seeing? Why is this happening? What caused this change in temperature?" [new ePCK for reflection in action and student outcomes during the albedo inquiry for Joy].

During data collection, we identified student issues regarding the shape of the heating curves. One student reported, "Sir, this [graph line] is curved." Many students expected the heating relationship to be a straight linear relationship; explaining this required reflection in action by Pat, copied by Joy, to remind students that a change in average kinetic energy is rarely linear [Pat's pPCK, Joy's new ePCK]. All the groups found the white fabric with the highest albedo, resulting in the lowest final temperature, while the black cloth had the lowest albedo and the highest final temperature. All 12 groups had predicted the black cloth would warm quickly, resulting in a higher temperature than the white fabric. This is not surprising based upon the amount of time we spent reinforcing, "White has the highest albedo." or by asking, "Which one reflects the most?" before the data were collected [pPCK]. It should be noted that many students enjoyed making the correct prediction.

5.3.2.3 Teach—The Ocean Acidification Inquiry

This enactment was the first teaching trial for the split ocean acidification inquiry and required extra preparation to gather local seawater, shells, and glacial ice (stored in a freezer). We started with the planned discrepant event, exposing the seashell in 1.0 M HCl. The students first made predictions that ranged from "Nothing will happen." to "The shell will totally dissolve!" Several students were surprised by the vigorous reaction, and most predicted either O_2 or CO_2 gas was evolving. When the burning stick was extinguished by immersion into the gas portion of the beaker, several students declared, "CO_2!" The second event was discrepant as it surprised many students that gas could be poured and extinguish a flame. One class asked for an encore after the candle was extinguished by the reaction's CO_2 product [ePCK reflection in action for student engagement and formative assessment]. These events started the simultaneous inquiries for seawater acidification and seashell solubility.

Day 1 Seawater Acidification Inquiry Group

After predicting what would happen, the students started bubbling their breath into 50 ml of seawater, resulting in a pH change made visible when the bromothymol blue indicator changed from blue to green and then yellow. The interface simultaneously graphed pH change, demonstrating how the seawater was initially basic and became acidic over time. The students reacted to the colour change, "Hey, it's going green!" We asked, "What does the colour change indicate? What is in your breath to cause this to happen?" The students set their yellow seawater solutions aside for the next day and researched the answers to the formative questions.

Day 1 Seashell Solubility Inquiry Group

After making predictions, the students' main tasks were to weigh the shells; place them in acidic, basic, and neutral solutions; make initial observations, and allow the reaction to occur overnight. The three solutions were the same colour, but two had unique odours. The basic solution was referred to as "clean-smelling" by one student, while another student noted the acidic solution "smelled disgusting," and the neutral solution had "no smell." The predictions for a reaction were divided between the bleach (base) and vinegar (acid) causing a reaction; no groups predicted the water would dissolve the seashells. The students started the reactions and observed them for 10 to 15 min and noted the bleach caused no reaction; the tap water caused a few bubbles to "grow" on the shell surface while the vinegar produced a steady stream of bubbles from the shell. Comments such as "This is not as much as your reaction." and "Ours is not bubbling like the first one." led to short discussions of acid strength and reactivity [ePCK, pedagogical knowledge reinforced by student outcomes, reflection in action for content knowledge].

Day 2 Collaborative Data Sharing by Laboratory Groups

In both classes, the data-sharing session started with a discovery that the acidified seawater turned blue overnight as the CO_2 gassed off and the pH of the seawater returned to above 7. Instantly, we regretted not leaving the pH probes in the solutions overnight to collect data but were pleased this predicted result occurred [reflection on action, reflection on student outcomes, ePCK]. The shells in vinegar (low pH) decreased in mass due to the loss of CO_2 during the chemical reaction, while mass changes in tap water were barely detectable. The students shared their pH data with the main findings that seawater pH changed quickly when exposed to CO_2 but reverted to its original status over time, and seashells were minimally soluble in tap water and weak acids.

A discussion of ocean acidification focused on information from a printed ocean acidification infographic that demonstrated the consumption of carbonate ions reacting with CO_2 could impact shell growth by impeding calcification (NOAA, 2021) [cPCK for ePCK of planning, teaching with reflection in action]. Following the discussions, the students collectively researched the post-laboratory summative questions.

5.3.2.4 Teach—Ocean Currents Inquiry

The six laboratory groups of each class combined into two large groups for the ocean currents inquiry. One group studied the model with alcohol thermometers, and the other used Vernier® temperature probes connected to the interface; an oxygen probe was added to examine any changes in dissolved oxygen created by currents. The heat lamp was placed outside, facing the water's surface opposite the ice. In a short review, many students recalled that cold water is denser than warm water and would sink in the water column. A class discussion earlier in the week on ocean currents included a review of the two major currents off our coast: the warm Gulf Stream from the south and the cold Labrador Current from the north. That class also discussed the forces that led to the formation of both currents: heat causing water expansion toward the poles from the equator, the falling of water at the poles due to cooling and increased density, and the narrowing and increased speed of currents along our coast due to the impact of the rotation of the Earth [reflection in action of new ePCK content knowledge from cPCK sources].

Before data collection, the students predicted what would happen in the model ocean due to the ice and the heat lamp. Student 1 stated, "After the light is turned on, the water will increase in temperature and go toward the iceberg ice and cool and then fall to the bottom." Student 4 hypothesized, "The warm water becomes cool and is forced downwards toward the cool water creating a current. Oxygen is pushed toward the bottom." It was clear from their predictions that the students were aware that cooling increased the density of water, causing it to sink toward the bottom; two students predicted oxygen levels at the bottom would increase due to this movement. Further, the aquarium ocean was a visual model to apply any of their knowledge and make predictions based upon the knowledge that cold water descends while warm water expands across the surface [ePCK—pedagogical reasoning for a new model]. After predictions, data collection started with a 5-min baseline without the heat lamp and then with 30 min of heating.

While the models produced many predictions, the aquarium oceans were not as dynamic as expected because we did not account for water's capacity to absorb heat from the lamp and to limit water expansion [reflection in action, new ePCK for pedagogical knowledge]. After 15 min, the temperatures changed slightly but not in the predicted pattern. One student commented, "Sir, this isn't doing very much!" We were concerned the inquiry would fail to demonstrate currents and result in a demonstration of water's high heat capacity. After 30 min, just before the class ended, Student 1 recorded, "Temperature near the lamp 9 °C and temperature at the bottom was 6 °C." This result supported the predicted modelling: cold water was sinking, and warm water was expanding and floating on the cooler water [ePCK pedagogical knowledge for the untested model of ocean currents]. Due to the extended data collection, the inquiry ended with no discussion of the data.

In the follow-up class, the data and laboratory report questions were discussed; it was noted that oxygen levels did not change (13.1 mg/L, both classes). Pat explained that the level of dissolved oxygen was high, almost full saturation, and would change little since the temperature at the bottom of the aquarium changed only slightly from 4–6 °C [pPCK content knowledge with reflection in action for the new model]. Both classes concluded the model ocean did not supply any visible evidence of a current flow, but conditions were present to create a current that we could not measure. To complete the report, students were asked to consider actual ocean currents to explain their observations.

The follow-up class continued with an examination of iceberg ice. The students were each given a piece of ice in a beaker and asked to observe it, touch it, then add water and listen. "It's fizzy!" exclaimed one of the students. "It pops!" said another. These observations led to the question, "Why does it pop?" Pat led a discussion of glaciers, and the formation of glacial ice based on Funder and Hansen's (1996) research [ePCK knowledge, reflection in action from cPCK]. We were pleasantly surprised that handling the ice had such an impact on the students, creating an unexpected discrepant event where we explained the popping and fizzing was compressed atmospheric gas being released from the ice [new reflection in action ePCK for us]. This led to students learning the Greenland Glacier is 2 km thick and calved icebergs were carried to Conception Bay by the Labrador Current. This reflection in action for ePCK of content knowledge was supported by an NSTA conference keynote presentation (Alley, 2015). The glacial ice was frozen for thousands of years, and the gas pockets in the iceberg were an atmospheric time capsule that allowed scientists to determine past CO_2 levels [cPCK that supported ePCK reflection in action for glacial ice]. Many students were fascinated by the glacial ice facts and seemed pleased they could use their senses to differentiate between freezer and glacial ice.

5.3.3 *Reflect—Student Outcomes on Summative Ocean Inquiry Reports*

Examining student outcomes in the laboratory reports allowed us **to witness student perspectives and learning during the activities and then reflect using pedagogical reasoning. Each set of reports was examined and reviewed with sample responses reported here from consenting students.**

5.3.3.1 Student Outcomes for the Albedo Activity

None of the albedo activity reports contained the previously encountered student misconception that the high albedo had the highest temperature change. The formative comments and questions before the inquiry may have addressed this misconception (perhaps too much). However, many student participants were confused by the report question: "In the summer, when the wind blows from the northwest across Conception Bay, the dark blue ocean with its low albedo does not significantly warm the air. Suggest two reasons why the air remains cool." One student's partial answer was, "The air remains cool because the ocean is a dark colour with a low albedo. So, all that heat is absorbed and not reflected back into the air." This response demonstrates an understanding of albedo but makes no mention of the heat capacity of water. Another student identified the other aspect of the correct answer, "It doesn't warm the air because the air is already cool because it comes from the north." Three students demonstrated a similar understanding of air masses; no report content clearly identified the heat capacity of the ocean water as a reason for the cool air; two reports did not submit an answer to this question. We concluded that more guidance using formative reminders of heat capacity and the moderating effect of large bodies of water would likely improve answers to this question [reflection on action, new ePCK for formative assessment]. Further, we believe these student outcomes demonstrate the complexity of local systems and the need to discuss local phenomena to help students address their daily experiences using an informed scientific perspective [reflection on student outcomes, new ePCK for knowledge of students].

5.3.3.2 Student Outcomes for the Ocean Acidification Inquiry

Unpacking the student predictions was difficult as many students misread the directions. In future, we would require answers to all these questions [reflection on action ePCK for pedagogical knowledge]:

1. What forms of carbon exist on the planet?
2. What processes add carbon to the atmosphere?
3. What processes remove carbon dioxide from the atmosphere?
4. What is the greenhouse effect?
5. How are humans connected to the carbon cycle?

Of the 14 reports examined, only three contained indications that seashells were a form of carbon; two students stated, "The formation of calcium carbonate removes carbon dioxide from [the] atmosphere." Many of the answers for Questions 3 to 5 were as expected, largely based on knowledge of climate change and a review of the carbon cycle; but several students seemed to struggle with the term *carbon* in Questions 1 and 2. During the inquiry, one student asked, "What do you mean by carbon, sir?" A similar question was posed by several groups and with a focus on Question 2. The students seemed to expect carbon dioxide versus carbon. Question 3 was a process question with carbon dioxide stated in the question along with the direction of the process; the answer *photosynthesis* was provided by all the students who were assigned that question [new ePCK based on student outcomes].

All 14 reports stated that carbon dioxide could be released from seashells, and 11 reports connected the increased carbon dioxide in the atmosphere with increasing ocean acidity. One student wrote, "The ongoing decrease in pH of the Earth's oceans is caused by the uptake of carbon dioxide from the atmosphere." Two report questions focused on the future, asking students to describe the impact of ocean acidification on plankton and sharks and, as an extension, "How will this impact us and our future, or what does it mean to you?" Nine students offered climate change strategies to help protect the future ocean. Interestingly, 11 students suggested ocean acidification will be a problem for sharks; however, these responses did not fit the expected connection of the shark population with the loss of plankton due to problems with shell production (identified in only 2 reports). In the report review, the impact of the loss of plankton was discussed in light of the ecology of food webs and energy pyramids where apex predators rely on planktonic primary production and first-order planktonic consumers [ePCK student misconceptions]. However, 10 students stated a more acidic ocean makes it difficult for sharks to smell (to detect food) while impacting swimming. We asked where the smell and swimming answer originated, and a student revealed he had Googled a journal for the answer and shared the reported findings of Rosa et al. (2017). These student outcomes resulted in reading the report and validating the student answers [new ePCK content knowledge from student outcomes].

The ocean acidification inquiry provided evidence of student connections with OLP 3. For example, one student stated, "It helped me understand how much carbon dioxide affects ocean acidification and the impact it will have on the earth." another student noted, "CO_2 causes acidity in the ocean." These comments echo most responses from the analyzed reports, demonstrating these students' understanding of ocean processes.

5.3.3.3 Student Outcomes for the Ocean Currents Inquiry

The students patiently waited during the ocean currents inquiry as changes to the system required 30 min. One student noted, "We replicated an ocean current with a lamp as the sun and heated the water and pushed toward the ice and cooled—the energy being transferred through advection and convection." Another student

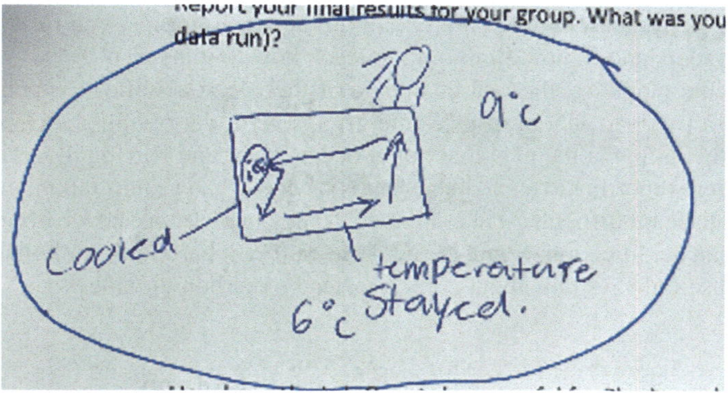

Fig. 5.2 Student model with proposed currents based on water temperature evidence

diagrammed a model of the movement of water (Fig. 5.2). Many students believed a current was created in the model ocean, with 9 suggesting the water expanded due to thermal expansion caused by the heat lamp; 11 students correlated cooling by ice with an increase in density, resulting in water sinking. Two students demonstrated a misconception or momentary confusion that heating would increase the density of water.

5.3.4 Reflect—Inquiry Activities

We both agreed that inquiry activities are crucial for student learning and that inquiry first, before the full review of theory, is a refreshing change. Joy stated, "It makes a connection with you, the kids, and the subject." The knowledge probes or predictions offered insights; Pat commented on students' lack of experience with the ocean, "They are raw with regards to doing science locally, talking about the ocean. Many of them were afraid of things that we were doing. They didn't know what to do with iceberg ice." Joy continued, "It's also very intimidating. I don't have the background. They don't have the background. They don't have the terminology yet. But if you give them the experience first with the activities, they had the connection to go further with it."

The ocean inquiry predictions provided explicit and useful student information. Joy noted, "You have the weak learner, you have the average learner, and then you have the one [who's] going to be gifted and excel. Through observation, you're able to determine who the learners are and how you can better meet their needs." Pat added, "We got to address it much more readily, like, for example, sometimes people making mistakes of pH … we got to address that with the ocean acidification activity." Joy continued, "Some of them understood it, but they couldn't physically put it down.

So, through conversation and conferencing in the small groups, you're able to say, 'OK, you understand it, now I'm going to teach how to transfer it'."

Despite the planning, the real challenge of the ocean acidification activity was that it was a two-part laboratory. "Some groups were dissolving shells, and other groups were looking at the pH [change] of ocean water and leaving it overnight and then they were reporting to each other. That's when we had groups talking to groups. That was a little bit difficult." Pat continued, "They knew about how carbon dioxide creates ocean acidification. Some of them are calling it carbon, I think that was my slang because I always talk about carbon instead of carbon dioxide."

5.3.5 *Reflect—Ocean Literacy*

Reflecting on our ocean literacy focus, Joy thought the students enjoyed learning about the chemical reactions of the ocean acidification inquiry, adding it was practical "because they could physically see what was happening as a result of the mixtures and a result of changing the pH balance." We found evidence of the students understanding that human activity leads to pollution and ocean acidification (OLP 6—D, NOAA, 2021, p. 11). Also, students understand that albedo changes can impact our climate and weather, and that radiation and cooling are fundamental for understanding ocean currents and energy flow (OLP 3—A & B, NOAA, 2021, p. 9).

5.4 Discussion

Science teachers' tacit knowledge, such as ePCK, is difficult to capture (Alonzo et al., 2019). However, our collaborative action research exploring PCK and ocean education activities included explicit verbal communication within the pedagogical reasoning of the plan–teach–reflect cycle (Carlson et al., 2019). As a result of this sharing during lesson creation and enactment, we found that ePCK development varied by phase of pedagogical reasoning (Table 5.1).

5.4.1 *The Plan–Teach–Reflect Cycle and the Development of PCK*

During planning, we shared pPCK for knowledge of student sensory and cognitive needs (Joy to Pat) and knowledge of curriculum PCK (Pat to Joy). For the albedo activity, we drew on our pPCK and used reflection on action, applying what we experienced in previous inquiries. However, the Plan phase of the two novel lessons had aspects of indetermination. Our pPCK was limited, and reflection on action resulted in pedagogical inferences of what may work as we had no student outcomes for the newly created discrepant events and the formative and summative assessments. Lesson enactment was the reflection in action context for new formative assessments and the application of new content knowledge for glacial ice (Funder & Hansen, 1996), ocean temperatures (Comiso, 2001), ocean current modelling (Reckling et al., 2014) and ocean acidification (Kelley et al., 2015). Nilsson and Vikström (2015) found that high school science teachers developing new lessons in learning study "changed their teaching of specific content in a way that provided new possibilities for students to grasp the objective learning" (p. 2852).

Some of our pedagogical inferences led to success, such as student data sharing of the ocean acidification activity; however, we regret not asking the students to collect data overnight to see the pH rebound of ocean water as the CO_2 gassed off. We were initially confident that the students would use existing group inquiry protocols for each new inquiry experience and that the new lesson predictions and collected data would result in a useful experience regardless of whether the laboratory activity proceeded as planned. This is evidenced by the student outcomes of the ocean currents inquiry (Fig. 5.2). To provide new possibilities for students, Pat changed the ocean currents inquiry the following year to include a dye visually demonstrating the current formation.

Even in remote high school contexts, prospective lesson ideas of the Plan phase of a lesson required student outcomes during enactment to be concretized as ePCK (Wells et al., 2023). Our pedagogical reasoning decisions for the novel lesson development and enactment demonstrate the importance of student outcomes for teacher learning. For us, planning was a dyadic, controlled interaction. Teaching was predominantly an individual activity and less controlled; once the inquiries started, we separated to manage, probe, assess, solve problems, answer questions, and collect feedback—these student outcomes were vital for our learning (Carlson et al., 2019). Our dyadic and individual teaching actions demonstrated the difference between reflecting on the action of planning and the reflecting in action of teaching. There was no action in planning, only thoughts of our past actions or projections of what may happen during instruction. During planning, we thought we prepared sensory-rich experiences; during enactment, the students provided sensory-rich signals to us along with cognitive challenges such as their questions, reactions, and misconceptions. These afferent signals informed our pedagogical reasoning when we drew on pPCK and ePCK at the moment to produce answers to questions, responses to reactions, and guidance to address misconceptions.

5.4.2 *Student as Agents of Ocean Learning*

Our ocean lessons addressed curriculum outcomes and OLPs 3 and 6 (i.e., the oceans' influence on weather and climate and the interconnectedness of humans with the ocean; Santoro et al., 2017). Further, engagement with local contexts was a vital "place" connection for the students (Arenas, 2001; Sobel, 2004).

The student groups were inquiry-proficient agents, but this was not always the case. When the groups formed months before the ocean inquiry activities, there was trepidation, similar to students' initial reluctance "to engage with unfamiliar and 'confusing' technologies," found by Hardy et al., (2019, p. 12). Experience and practice with task-related feedback changed confusion into routine (Hattie & Clarke, 2018). The application of data collection skills during the ocean inquiry activities was evidence of the rise of student agency, similar to that reported by Hardy et al. (2019). Furthermore, the document review was a rich source of student thinking and taught us that students were researching the answers to our questions, such as the impact of ocean acidification on sharks, and were finding unexpected answers in the ocean science literature. The practice of inquiry, with hands-on, engaging social learning (Wilson et al., 2010), had taken the students beyond lesson content to generate their own knowledge. We learned how the lessons engaged students' senses, from the popping iceberg ice to coloured visualizations of pH change and chemical reactions connected with scent. These sensory components increased accessibility in ocean learning as indicated by Carp (2007) and supported students with IEPs who Clarke & Kast (2020) suggested may benefit more from inquiry-based science instruction.

5.5 Concluding Remarks

A limitation of this study was the lack of student interviews. While observations and the document review were fruitful, interviews would have allowed us to ask more specific questions about student learning and report student connections and commentary for lesson strategies, such as the ocean acidification inquiry student reporting. Many students indicated they enjoyed the ocean lessons but, without specific quotes, the power of their comments is not passed on to the reader.

The interaction of the types of PCK, ePCK, pPCK, and cPCK, are pathways of skill and knowledge that vary by teacher and context (Alonzo et al., 2019) but are tracible (Fig. 5.1 and Table 5.1). However, examining the pedagogical reasoning of our dyad was complicated; a web of interactions that plays into pedagogical decisions with ePCK use and development in the context of its expression.

We believe the knowledge gains from planning are prospective, reflection on action, and not yet pPCK. When teaching, the active enactment of planning and corresponding student actions and ocean lesson student outcomes are the data that is reflectively processed through a teacher's pedagogical reasoning (Carlson et al., 2019). After a lesson enactment, what ePCK remains? What worked in action will be repeated, what requires modification will be changed, and that which had no impact may be discarded. The contexts of planning-, teaching-, and reflection-related pedagogical decisions resulted in ePCK recruitment into our topic-specific pPCK. Ultimately, our reasoning for each stage, combined with the student outcomes, afforded us the opportunity for high-quality teacher learning and the development of new pPCK.

Acknowledgements This research was supported by the English School District of Newfoundland and Labrador (ESDNL).

References

Adu-Boateng, S., & Goodnough, K. (2022). Examining a science teacher's instructional practices in the adoption of inclusive pedagogy: A qualitative case study. *Journal of Science Teacher Education, 33*(3), 303–325. https://doi.org/10.1080/1046560X.2021.1915605

Alley, R. B. (2015, November 12–14). *Good news on energy and the environment: The fruits of STEM education* [Keynote address]. NSTA Area Conference on Science Education. Philadelphia, PA, United States.

Alonzo, A. C., Berry, A., & Nilsson, P. (2019). Unpacking the complexity of science teachers' PCK in action: Enacted and personal PCK. In A. Hume, R. Cooper, & A. Borowski (Eds.), *Repositioning pedagogical content knowledge in teachers' knowledge for teaching science* (pp. 273–288). Springer. https://doi.org/10.1007/978-981-13-5898-2_12

Arenas, A. (2001). 'If we all go global, what happens to the local?' In defence of a pedagogy of place. *Educational Practice & Theory, 23*(2), 29–47. https://doi.org/10.7459/ept/23.2.03

Banchi, H., & Bell, R. (2008). The many levels of inquiry. *Science & Children, 46*(2), 26–29.

Banfield, C. E. (1993). The climate of Newfoundland. In A. Robertson, S. Porter, & G. Brodie (Eds.), *Climate and weather of Newfoundland and Labrador* (pp. 13–31). Creative Publishers, St. John's.

Bell, T., Briggs, R., Bachmayer, R., & Li, S. (2014). Augmenting Inuit knowledge for safe sea-ice travel—The SmartICE information system. *IEEE 2014 Oceans—St. John's Proceedings,* 1–9. https://doi.org/10.1109/OCEANS.2014.7003290

Carlson, J., Daehler, K. R., Alonzo, A. C., Barendsen, E., Berry, A., Borowski, A., Carpendale, J., Kam Ho Chan, K., Cooper, R., Friedrichsen, P., Gess-Newsome, J., Henze-Rietveld, I., Hume, A., Kirschner, S., Liepertz, S., Loughran, J., Mavhunga, E., Neumann, K., Nilsson, P., Park, S., ... Wilson, C. D. (2019). The refined consensus model of pedagogical content knowledge in science education. In A. Hume, R. Cooper, & A. Borowski (Eds.), *Repositioning pedagogical content knowledge in teachers' knowledge for teaching science* (pp. 77–94). Springer. https://doi.org/10.1007/978-981-13-5898-2_2

Carp, M. (2007). Seeing things in a whole new way: Working with visually impaired children. *Current: The Journal of Marine Education, 23*(2), 23–24.

Clarke, L., & Kast, D. J. (2020). A case study in making ocean education accessible for students with special needs. *Current: The Journal of Marine Education, 34*(2), 28–35. https://doi.org/10.5334/cjme.29

Comiso, J. C. (2001). Satellite-observed variability and trend in sea-ice extent, surface temperature, albedo and clouds in the Arctic. *Annals of Glaciology, 33*, 457–473. https://doi.org/10.3189/172756401781818617

Creswell, J. W. (2015). *Educational research: Planning, conducting, and evaluating quantitative and qualitative research* (5th ed.). Pearson.

Dana, N. F., & Yendol-Hoppey, D. (2014). *The reflective educator's guide to classroom research: Learning to teach and teaching to learn through practitioner inquiry* (3rd ed.). Corwin Press.

Erickson, B., & Crews, T. (2019). From dissolution to solution: New approaches to teaching ocean acidification. *The Science Teacher, 86*(5), 56–63. https://doi.org/10.2505/4/tst18_086_05_56

Fauville, G. (2018). Ocean literacy in the twenty-first century. In G. Fauville, D. L. Payne, M. E. Marrero, A. Lantz-Andersson, & F. Crouch (Eds.), *Exemplary practices in marine science education: A resource for practitioners and researchers* (pp. 3–11). Springer. https://doi.org/10.1007/978-3-319-90778-9_1

Fauville, G., Payne, D. L., Marrero, M. E., Lantz-Andersson, A., & Crouch, F. (2018). Exemplary practices in marine science education: A resource for practitioners and researchers. *Springer*. https://doi.org/10.1007/978-3-319-90778-9

Florian, L., & Beaton, M. (2018). Inclusive pedagogy in action: Getting it right for every child. *International Journal of Inclusive Education, 22*(8), 870–884. https://doi.org/10.1080/13603116.2017.1412513

Funder, S., & Hansen, L. (1996). The Greenland ice sheet—A model for its culmination and decay during and after the last glacial maximum. *Bulletin of the Geological Society of Denmark, 42*(2), 137–152. https://doi.org/10.37570/bgsd-1995-42-12

Gess-Newsome, J. (2015). A model of teacher professional knowledge and skill including PCK: Results of the thinking from the PCK summit. In A. Berry, P. Friedrichsen, & J. Loughran (Eds.), *Re-examining pedagogical content knowledge in science education* (pp. 28–42). Routledge. https://doi.org/10.4324/9781315735665

Glithero, L. (2020, June). *Understanding ocean literacy in Canada: National report.* Canadian Ocean Literacy Coalition. http://colcoalition.ca/wp-content/uploads/2021/02/COLCNationalReportFINAL__.pdf

Goodnough, K. (2010). Teacher learning and collaborative action research: Generating a "knowledge-of-practice" in the context of science education. *Journal of Science Teacher Education, 21*(8), 917–935. https://doi.org/10.1007/s10972-010-9215-y

Gorospe, K. D., Fox, B. K., Haverkort-Yeh, R. D., Tamaru, C. S., & Rivera, M. A. J. (2013). Engaging students in the Pacific and beyond using an inquiry-based lesson in ocean acidification. *Journal of Geoscience Education, 61*(4), 396–404. https://doi.org/10.5408/12-390.1

Guest, H., Lotze, H. K., & Wallace, D. (2015). Youth and the sea: Ocean literacy in Nova Scotia, Canada. *Marine Policy, 58*, 98–107.

Hattie, J., & Clarke, S. (2018). *Visible learning: Feedback*. Routledge.

Han, G., Ma, Z., & Chen, N. (2019). Ocean climate variability off Newfoundland and Labrador over 1979–2010: A modelling approach. *Ocean Modelling, 144*, 101505. https://doi.org/10.1016/j.ocemod.2019.101505

Hardy, L., Dixon, C., & Hsi, S. (2019). From data collectors to data producers: Shifting students' relationship to data. *Journal of the Learning Sciences, 29*(1), 1–23. https://doi.org/10.1080/10508406.2019.1678164

Hodson, D. (2014). Learning science, learning about science, doing science: Different goals demand different learning methods. *International Journal of Science Education, 36*(15), 2534–2553. https://doi.org/10.1080/09500693.2014.899722

Hofstein, A., & Lunetta, V. N. (2004). The laboratory in science education: Foundations for the twenty-first century. *Science Education, 88*(1), 28–54. https://doi.org/10.1002/sce.10106

Hoover, T. (2017). Disequilibrium: Wildfires, the fire triangle, and CO_2 extinguishers. *Science Scope, 41*(4), 14–17. https://doi.org/10.2505/4/ss17_041_04_14

Kelley, A. L., Hanson, P. R., & Kelley, S. A. (2015). Demonstrating the effects of ocean acidification on marine organisms to support climate change understanding. *American Biology Teacher, 77*(4), 258–263. https://doi.org/10.1525/abt.2015.77.4.5

Kelly, R., Elsler, L. G., Polejack, A., van der Linden, S., Tönnesson, K., Schoedinger, S. E., ... & Wisz, M. S. (2022). Empowering young people with climate and ocean science: Five strategies for adults to consider. *One Earth, 5*(8), 861-874.

Liboiron, M. (2021). Pollution is colonialism. *Duke University Press*. https://doi.org/10.1215/9781478021445

Monroe, M. C., Plate, R. R., Oxarart, A., Bowers, A., & Chaves, W. A. (2019). Identifying effective climate change education strategies: A systematic review of the research. *Environmental Education Research, 25*(6), 791–812.

National Oceanic & Atmospheric Administration. (2021). *Ocean literacy: The essential principles and fundamental concepts of ocean sciences for learners of all ages* (Version 3.1). https://oceanservice.noaa.gov/education/literacy/OceanLiteracy_ENG_V3.1_2021-1.pdf

National Research Council. (2012). A framework for K–12 science education: Practices, crosscutting concepts, and core ideas. National Academies Press. https://doi.org/10.17226/13165

National Academies Press. https://doi.org/10.17226/13165Nilsson, P., & Vikström, A. (2015). Making PCK explicit—Capturing science teachers' pedagogical content knowledge (PCK) in the science classroom. *International Journal of Science Education, 37*(17), 2836–2857. https://doi.org/10.1080/09500693.2015.1106614

Nilsson, P., & Vikström, A. (2015). Making PCK explicit—Capturing science teachers' pedagogical content knowledge (PCK) in the science classroom. *International Journal of Science Education, 37*(17), 2836–2857. https://doi.org/10.1080/09500693.2015.1106614

Putland, J., Hoeberechts, M., Pelz, M., Hudson, L., Tolmie, C., & Carrasquilla, M. (2021). Including the ocean in formal K-12 climate education: Assessment of a lesson for middle and high school students. *Canadian Journal of Environmental Education (CJEE), 24*(1), 189–212.

Reckinger, S. M., Aca, B., & Pitz, K. (2014, April). Ocean circulation in a rotating tank: An outreach project in fluid dynamics. *IEEE Proceedings of the 2014 Zone 1 Conference of the American Society for Engineering Education*, 1–6. https://doi.org/10.1109/ASEEZone1.2014.6820678

Rosa, R., Rummer, J. L., & Munday, P. L. (2017). Biological responses of sharks to ocean acidification. *Biology Letters, 13*(3), 20160796. https://doi.org/10.1098/rsbl.2016.0796

Rose, G. A. (Ed.). (2018). *Atlantic cod: A bio-ecology*. Wiley. https://doi.org/10.1002/9781119460701

Sagan, C., Toon, O. B., & Pollack, J. B. (1979). Anthropogenic albedo changes and the Earth's climate. *Science, 206*(4425), 1363–1368. https://doi.org/10.1126/science.206.4425.1363

Santoro, F., Santin, S., Scowcroft, G., Fauville, G., & Tuddenham, P. (Eds.). (2017). *Ocean literacy for all—A toolkit*. IOC/UNESCO & UNESCO. https://unesdoc.unesco.org/ark:/48223/pf0000260721

Saunders-Stewart, K. S., Gyles, P. D. T., & Shore, B. M. (2012). Student outcomes in inquiry instruction: A literature-derived inventory. *Journal of Advanced Academics, 23*(1), 5–31. https://doi.org/10.1177/1932202x11429860

Shulman, L. S. (1987). Knowledge and teaching: Foundations of the new reform. *Harvard Educational Review, 57*(1), 1–23. https://doi.org/10.17763/haer.57.1.j463w79r56455411

Smith, G. A. (2002). Place-based education: Learning to be where we are. *Phi Delta Kappan, 83*(8), 584–594. https://doi.org/10.1177/003172170208300806

Sobel, D. (2004). *Place-based education: Connecting classrooms and communities*. Orion Society.

Stewart, A. (2021). What is Canadian ocean literacy? A community perspective essay [Editorial]. *Canadian Journal of Environmental Education, 24*(1), 11–18.

Strang, C., deCharon, A., & Schoedinger, S. (2007). Can you be science literate without being ocean literate? *Current: The Journal of Marine Education, 23*(1), 7–9.

Tran, L. U., Payne, D. L., & Whitley, L. (2010). Research on learning and teaching ocean and aquatic sciences. *NMEA Special Report, 3*(1), 22–26.

Wells, P. R., Boyce, R., Gregory, L., Locke, J., Halleran, S., & Woodland, M. (2017). Mobile devices—Friend or foe? *The Bulletin, 60*(5), 26–27.

Wells, P. R., Goodnough, K., Azam, S., & Galway, G. (2023). Changes in high school distance education science teachers' pedagogical content knowledge (PCK) during remote lesson study. In R. Huang, N. Helgevold, J. Lang, & H. Jiang (Eds.), *Teacher professional learning through lesson study in virtual and hybrid environments: Opportunities, challenges, and future directions* (pp. 203–222). Routledge. https://doi.org/10.4324/9781003286172

Wells, P. R., & Negrijn, J. (2019). Lessons from monitoring ocean water temperatures. *Journal of Ocean Technology, 14*(2), 57–64.

Wilson, C. D., Taylor, J. A., Kowalski, S. M., & Carlson, J. (2010). The relative effects and equity of inquiry-based and commonplace science teaching on students' knowledge, reasoning, and argumentation. *Journal of Research in Science Teaching, 47*(3), 276–301.

Worm, B., Elliff, C., Fonseca, J. G., Gell, F. R., Serra-Gonçalves, C., Helder, N. K., Murray, K., Peckham, H., Prelovec, L., & Sink, K. (2021). Making ocean literacy inclusive and accessible. *Ethics in Science & Environmental Politics, 21*, 1–9. https://doi.org/10.3354/esep00196

Chapter 6
Providing Feedback on Student Laboratory Reports: Issues in the Formative Assessment Practices of Preservice Science Teachers

G. Michael Bowen, Anthony W. Bartley, and Wayne Melville

6.1 The Function of Laboratory Reports

Perhaps the quintessential activity that distinguishes science classes from any other class in high school are the student investigations and experiments and the subsequent student-produced laboratory reports that detail practices and findings in an investigation (Gericke et al., 2023). Laboratory reports are a particular genre of science writing considered preparatory practice for both undergraduate studies and formal research reports in science courses (Parkinson, 2017). These documents have a quasi-standardized structure seemingly modeled on research reports found in professional science journals (Bradley, 2005). Traditionally, the standard high school laboratory report follows a format that includes sections with a hypothesis or focus question, an overview of what is known (akin to a literature review), a description of needed materials and the data collection procedure, a presentation of the data (usually including tables, graphs, and written descriptions), and conclusions (including discussion and implications; Gericke et al., 2023; Lunetta, 1998; Woolnough & Alsop, 1985).

Laboratory reports serve several key functions in science education. They provide students the opportunity to engage in the full scientific process: From forming hypotheses to collecting and analyzing data to drawing conclusions (Hofstein &

G. Michael Bowen (✉)
Faculty of Education, Mount Saint Vincent University, Halifax, NS, Canada
e-mail: gmbowen@yahoo.com

A. W. Bartley · W. Melville
Lakehead University, Thunder Bay, ON, Canada
e-mail: abartley@lakeheadu.ca

W. Melville
e-mail: wmelvill@lakeheadu.ca

C. D. Tippett and T. M. Milford (eds.), *Seeing Science Through the Eyes of Canadian Teachers and Learners*, Contemporary Trends and Issues in Science Education 54,
https://doi.org/10.1007/978-3-031-88785-7_6

Lunetta, 2004; Wilcox & Lewandowski, 2016). This hands-on experience helps reinforce scientific concepts and methods. Laboratory reports allow students to practice critical scientific communication skills by organizing information and clearly presenting it in a standardized format (Volkmann & Abell, 2003). The laboratory report serves as evidence of student learning and can be assessed by instructors (Deters, 2005). Overall, laboratory reports can be seen as evidence of a crucial component of science curricula/learning, bridging the gap between scientific theory and practice.

6.1.1 The Role of Writing Laboratory Reports

An effective process for high school students is to present their inquiry investigation findings to a broader audience that involves receiving feedback from an external reviewer on how to improve their report. However, Canadian teacher education programs often lack courses that specifically teach assessment approaches tailored to evaluating discipline-specific work such as assessing the quality of science laboratory reports. Developing assessment literacy around these specialized formats could better prepare preservice science teachers (PSTs) to provide meaningful feedback and guide their students in effectively communicating their scientific investigations.

6.1.2 Feedback as Formative Assessment

What is essential is that any meaningful feedback should evoke thoughtful reflection in which all students can be encouraged to take part; only then can the formative process begin. In short, the dialogue between students and a teacher should be thoughtful, reflective, focused to evoke and explore understanding, and conducted so that all students have an opportunity to think and express their ideas. Pre- and inservice teachers should be provided with formative assessment experiences as it is unlikely that they will have had many encounters with providing formative feedback on laboratory reports, either from their own teachers (when they were K-12 students) or any of their university instructors (Atjonen et al., 2022; Black & Wiliam, 1998a; Black et al., 2003; Dassa et al., 1993; Daws & Singh, 1996; Staberg et al., 2023).

In our thinking for designing this study, we focused on the following suggestions from Black et al. (2004) that were embedded into the guidance for evaluating laboratory reports that we provided for our participants:

- Written tasks, alongside oral questioning, should encourage students to develop and show understanding of the key features of what they have learned.
- Comments should identify what has been done well and what still needs improvement and give guidance on how to make that improvement.

- Opportunities for students to respond to comments should be planned as part of the overall learning process.

In settings where teachers are evaluating their students' laboratory reports, their feedback can be used not only as formative assessment by the students in writing later reports (or improving the current one) but also by the teacher when that information is used to adapt later teaching and guidance on activities to better meet their students' needs with respect to eliciting the writing outcomes desired from the classroom activity (Shavelson et al., 2008). However, in this study reviewing these reports on high school student inquiry investigations, the PSTs were asked to provide constructive feedback for the student-authors that could encourage them to improve their current report—but which their teachers could not use in a formative sense because they had not instructed these students—and that they could use to guide their future teaching practices.

6.1.3 Preservice Science Teachers and Assessment Practices

PSTs often have limited exposure to authentic inquiry investigations during their own education (Bowen & Bartley, 2007). This limitation can present challenges when it comes to assessing high school laboratory reports. Research suggests that PSTs may benefit from practice providing formative assessment on laboratory reports (Staberg et al., 2023). Bowen and Bartley (2007) found that many PSTs lacked the necessary skills and experience to effectively evaluate student laboratory work and that they tended to focus more on the procedural aspects of laboratory reports rather than deeper conceptual understanding, which is similar to findings by Zemel et al. (2021). This might occur because PST practices during practicum often focus more on those teaching practices than on the type and manner of feedback the PSTs might provide their students (Ruelmann et al., 2023).

Providing PSTs with opportunities to assess authentic student laboratory reports has been proposed to develop their assessment skills and could occur outside of the practicum experience (Bowen & Bartley, 2007; Espiritu et al., 2018; Luft, 2001). This hands-on practice would allow PSTs to experience the complexities of evaluating open-ended investigative work, which can differ significantly from the more structured laboratory activities they may have encountered as students themselves.

6.1.4 Resources and Guidelines for Assessing Laboratory Reports

Publication of the American *Next Generation Science Standards* (National Research Council, 2013) led to the development and publication of resources supporting teacher assessment of student investigations; for example, *Developing Assessments for the Next Generation Science Standard* (National Research Council, 2014). Several online resources provide checklists/rubrics for component parts of laboratory reports, but these items are usually considerably open to interpretation and provide little guidance on how to use them. Overwhelmingly, these online resources were intended to help Teaching Assistants evaluate undergraduate laboratory reports, which seemed to be strongly modeled on professional science publications. There appeared to be (a) few resources intended to effectively guide PSTs who want to understand how to evaluate and provide feedback on laboratory reports and (b) few academic resources that describe and conceptualize those practices. More importantly, the resources that do exist implicitly assume a level of understanding about science evidence claims found in scientific investigations exceeding that which research, and our experience, suggests that PSTs hold. There are "very few studies [that] have closely examined the type of feedback that pre-service teachers deliver when assessing written work" for any subject (Máñez et al., 2024, p. 243) not to mention for laboratory reports.

In this study, we wanted to better understand the details of how PSTs provide feedback on high school students' written laboratory reports. We developed an online, peer-reviewed publishing site for high school inquiry projects, with the idea that the review of articles submitted to this online journal would initially be conducted by PSTs at universities across Canada. We had two intentions with this activity: (a) to provide PSTs experiences with an inquiry investigation activity that is central to the conduct and progress of professional science, modelling a way in which high school students can experience such an approach; and (b) to provide PSTs with some applied experience with an authentic formative assessment activity. This chapter provides a detailed examination and analysis of the reviewing/formative feedback practices engaged in by the participating PSTs.

6.2 Research Design

This study built on a previous study (Bowen & Bartley, 2007) where secondary science education students (all having BSc degrees) in an advanced science methods course as part of a 2-year post-baccalaureate BEd program wrote reflective commentaries on their experience providing formative feedback on a single Grade 12 laboratory report. The initial study suggested that the PSTs had insufficient resources (e.g., experiential, material) to provide effective feedback. The study presented in this chapter investigated the types of feedback PSTs provided on laboratory reports with the support of researcher-designed resources (Appendices A & B).

Four PSTs at an eastern Canadian university volunteered to participate in this follow-up study. Data were collected on the PSTs' formative assessment of three student laboratory reports in three phases, with each PST individually examining a different laboratory report in each phase. In other words, each PST had examined and provided written formative feedback on all three laboratory reports by the end of the final phase. The three different laboratory reports were from open-inquiry investigations and individually written by Grade 12 biology students. These reports were written toward the end of the school year and in a school that used open inquiry and problem-based learning approaches to science in Grades 9–12 (Jones et al., 2009; Melville et al., 2014). Each report had issues with representation of data in tables and graphs and the strength of the claims being drawn. The PSTs were instructed: "To provide feedback that would allow the paper to be the 'best paper possible,' but you cannot require them to re-collect data. You need to guide them to draw claims that represent the best possible from the data they collected."

In the first phase of the study, PSTs were asked to review one of three randomly assigned laboratory reports and submit written formative feedback for the student-author. There was no restriction placed on the time they could spend or what other resources (e.g., library or internet) they could use. In the second phase, PSTs were individually video recorded during a 90 to 120-min formative review of a laboratory report (randomly assigned from the remaining two possibilities) during which they used a think-out-loud approach and discussed what they were reading/thinking/doing/writing as their written review was being constructed. The PSTs had at-hand an internet-connected computer to search for needed information while conducting their review and to write their formative feedback. Following this video-taped activity, PSTs were each engaged in a short debriefing interview and provided a 2-page hint sheet (Appendix B) for conducting reviews of laboratory reports, which they were encouraged to examine before the third phase. This handout on how to improve reviewing of laboratory reports was constructed based upon insights gained from the first study (Bowen & Bartley, 2007) and from our informal review of resources for reviewing for professional journals and grading undergraduate laboratory reports. Other than use of the hint sheet, the conditions for the third phase were the same as for the second phase.

Our analyses of the surveys, transcripts, and reflective commentaries were based on interaction analysis (Jordan & Henderson, 1995) and grounded theory (Strauss & Corbin, 1990). In our examination of the dataset, we independently viewed the videos and read the transcripts, forming individual assertions about potential themes that were compiled in draft analytic notes. In joint discussion sessions, we presented our individual tentative assertions, discussed our observations/tentative conclusions to reach consensus, and examined the dataset for confirming or disconfirming evidence before arriving at an agreed-upon interpretation of the data. Our final conclusions arose from many such iterations.

6.3 Findings

A variety of resources were analyzed for this study including the original high school student reports and transcripts/text for each type of feedback provided by the PSTs. We first discuss the high school student reports, how they were structured, and any issues identified in the reports. An analysis of the written formative feedback provided by the PST reviewers for the students is then presented and contextualized using the audio/video records compiled while the formative reviews were being constructed.

6.3.1 Analysis of High School Student Laboratory Reports

The student laboratory reports were constructed in a traditional report fashion (as described previously). All laboratory reports dealt with a specific aquatic plant, *Bacopa monnieri*, and dealt with a student-selected variable that would influence the photosynthetic rate of that plant. For instance, one laboratory report was entitled "The Effect of Nutrient Concentration on the Photosynthetic Rate of *Bacopa monnieri*." All laboratory reports were structured in a common manner with an Introduction, a Hypothesis Statement, a Materials list, a step-by-step Methods section (Report A had a diagram of the experimental set-up), a Results section with a table and a graph, a Discussion section, a Conclusion/Summary Statement section, and a References Cited list. This similarity of structure suggests that students had a template to standardize their laboratory reporting practices.

From a review conducted by this chapter's authors it was determined that all three student laboratory reports had issues in (a) data table structure, (b) graph structure, and (c) the claims and implications drawn from the findings. As opposed to tables typically found in science reports that afford visual inspection for trends, two of the three laboratory reports contained tables that were constructed in such a way as to confound this possibility (e.g., using four separate simple tables rather than one comprehensive table). None of the tables had rows/columns representing the average outcome despite replicate treatments. Each of the three laboratory reports contained a graph representing the major findings of the study; two reports used a bar graph and one report used a variant of a line graph with a numeric x-axis but with differences in temperature having identical physical spacing as would be found in a bar or line graph but with curvilinear fitted lines passing through the data point (see Appendix A).

Consistent with normally enacted practices in science, a canonical graphical representation of this dataset would have represented the data in a scatterplot graph in all three laboratory reports (Bowen & Bartley, 2013). Other structural issues were also apparent. One laboratory report had the abscissa and line variables reversed with curvilinear lines of best fit drawn. One bar graph mislabeled the abscissa. The line graph had lines of best fit through individual replicates instead of through the average. Axis labels were absent in the two bar graphs, and all three graphs had titles that were generic enough that they contributed to difficulties in understanding the patterns shown in the graphs. None of the results sections in any of the laboratory reports had any written description of the trends apparent in either the tables or the graphs.

6.3.2 Analysis of PSTs' Written Formative Feedback on Laboratory Reports

In each phase of the study, PSTs were instructed to provide formative feedback so that the high school students could respond to the feedback to produce the best report possible but without redoing any research activity. However, PST reviewers were instructed that making suggestions on methodological or sampling grounds that could be redressed in a future investigation activity was appropriate.

6.3.2.1 Review Phase I—Written Formative Feedback of 1st Laboratory Report

The first reviewing phase involved PSTs engaging in providing written formative feedback of one laboratory report. First, they read a paper on formative assessment (Black & Wiliam, 1998b) that was followed by a brief discussion in class. PSTs downloaded a laboratory report then completed and submitted their formative review.

All four PST reviewers made comments regarding the structure and style of the laboratory report (e.g., whether a particular comment should be in the methods or the introduction, how statements were made in the report, clarification of use of terms such as size). Two of the four reviewers suggested that a diagram of apparatus set-up would have helped them better understand and critique the study; only one laboratory report contained such a diagram. All four reviewers suggested that a written description of patterns in the data was warranted.

Regarding visual representations, one reviewer commented that "the author did a very good job of presenting his results" despite the laboratory report including an incorrect graphical representation and a data table lacking an average column. Another reviewer, reviewing the laboratory report that had four separate two-by-five tables, suggested "possibly displaying all this information in one table [would] make it easier for readers to compare values between different treatments." The same reviewer also commented that "the type and arrangement of the graph you have selected is a great way to display all of the results in one section" despite what we noted as a presentation that made interpretation difficult. Neither of the two other reviewers commented on the graphical representations needing changes other than benefitting from more specific axes labels.

Laboratory reports discussed the amount of oxygen production but did not discuss the rate of photosynthesis. Only one reviewer commented on the omission: "You have presented data to show the amount of oxygen produced but not really the calculated rate of production. Just be careful referring to 'rate' unless you are calculating and expressing it as a rate (e.g., amount/time)."

6.3.2.2 Review Phase II—Think-Out-Loud Review of 2nd Laboratory Report

In the second review phase, the four PSTs were video recorded while they used a think-out-loud approach to conducting a formative review of a different laboratory report. All Phase II reviews were longer than the Phase I reviews despite there being no additional instruction or support regarding providing formative feedback. In this review, similar comments about structure and style were made. For example, reviewers identified a need for more detail and clarity in the methods section. Three reviewers mentioned the need for a diagram of the apparatus to help readers understand how the inquiry was done. One reviewer wrote a detailed explanation as to why a diagram was needed: "someone reading your paper should be able to repeat your experiment. Most experiments that I have attempted to replicate are exponentially clearer when they include a diagram of the setup. Detail and scale are usually not that important, but a rough visual aid is indispensable." All reviews contained positive feedback for what was written in the laboratory reports.

6.3.2.3 Review Phase III—Think-Out-Loud Review of 3rd Laboratory Report with Hint Sheet

The third phase was similar to the second phase except that only three PSTs participated. The single modification was that the PSTs were given the handout "Suggestions for Reviewing a Laboratory Report" (Appendix B) to read before the final review. In general, the same pattern of positive feedback was noted with all three reviewers; and all Phase III reviews were longer than the Phase I and II reviews. Two reviewers identified a need for more details and clarification in the method, although two reviewers praised the student for including a diagram of the apparatus set-up. Only one reviewer indicated a need for a different type of graph; this same reviewer identified a need for an average column in the table. One reviewer provided only positive feedback for the table, and the other reviewer made no comments about the tables or graph in their report.

6.3.3 Overview of Formative Reviewing Practices

The PST reviewers made positive comments on many aspects of the laboratory reports. Our own analysis of the reports supports some of the positive feedback but suggests that some of the positive feedback from the PSTs was inaccurate. Of particular concern was the lack of critical feedback regarding visual representations (e.g., table structure, choice of appropriate graphs). This omission is important given the increase in length of the reviews over the three reviewing phases, which highlights that the PST reviewers were providing more and more detailed feedback on issues and yet did not provide needed feedback on visual representational practices. This finding is particularly noteworthy given that the constraints of the review allowed asking for changes; for example, a visual representation change would require only a re-presentation of collected data. The laboratory reports contained data, but the students did not necessarily analyze it. Several PSTs indicated that the use of simple statistics for analysis purposes would have been appropriate.

The titles and stated hypotheses of all three laboratory reports included factors that would affect the rate of photosynthesis. However, rate was unexplored; instead, laboratory reports dealt with amounts of gas production. Where time between data measurements is constant and the value recorded represents a net change, then a bar or line graph where the abscissa values are equidistant would be consistent with the concept of rate. However, in these laboratory reports, the graphs violated one or both conditions; thus, photosynthetic rate was not examined. Despite the centrality of rate to the stated objectives of each laboratory report, only one PST's feedback, during Phase I, addressed its absence; only one other PST addressed the issue during the think-out-loud but then did not mention it in the written formative feedback.

In fact, concerns that PSTs expressed during the think-out-loud were often not included in the written feedback, at least in detail sufficient for the student authors to be able to act on them. For example, a statement such as "This would make it known to you what relationships are present." is arguably lacking in sufficient detail to help a student. It should be noted that PSTs infrequently offered any specific advice to the students regarding how to provide a better analysis. This lack is particularly concerning given that there were issues with all graphs and an absence of discussion on rate that could have been commented upon.

The hint sheet provided for the third phase was used by only one PST as a checklist to structure their feedback, which only minimally highlighted needs for changes in the tables and graphs. They commented,

> Now to some extent when I marked the last paper, I did a number of these things. But I think as somebody marking a paper, it's a really helpful thing to have some sort of brief guide in front of you that you can use as a reference while marking, kind of a checklist sort of thing and a way of organizing your thoughts. I think it would not only make the feedback you give more thorough, but it would be more consistent between students, and I think to a large extent it would also be more efficient.

Two PSTs read the provided hint sheet at the beginning of the video recording, set the sheet down, and then did not appear to refer to it again during the review process.

6.4 Implications and Conclusions

Overall, we would claim that the PSTs in this study provided reasonable formative feedback on the student laboratory reports in many ways; however, there were areas that were cause for concern. These problematic areas dealt almost exclusively with the use of tables and graphs in ways that would afford discussion of the data in relation to the hypothesis (in this case, focusing on issues conflating the concept of rate and amount). These areas have previously been identified as being problematic for PSTs (Bowen & Roth, 2005; Bowen et al., 2016).

The PSTs seemed to have had difficulty providing the type of formative feedback on the laboratory reports that would have had the student authors engage in canonical reporting practices. While the four PSTs had not taken courses on assessment and had limited experience with authentic student work, this factor does not explain the difficulty they had with providing feedback on visual representations and data analysis. Previous research suggests that PSTs with BSc degrees have considerable difficulty with both interpreting (ÇİL & Kar, 2015; Roth et al., 1998) and using (Bowen & Roth, 2005; Bowen et al., 2016; Melear, 1999) visual representations in their own work, which is likely related to the difficulty our participating PSTs had in providing formative assessment on these laboratory reports.

The insights gleaned from this study suggest that there are issues with preparing PSTs for assessing student laboratory reports. Overall, the time PSTs spent on each report using the think-out-loud approach far exceeded what would be spent on typical grading in an actual classroom. However, these PSTs had more time for reflection, thought, interpretation, and analysis of student work when providing feedback, which suggests that even less critical feedback would be provided when assessing in a real-world timeframe. Laboratory reports are a formalized form of communicating science findings, but their fundamentals (e.g., data summarization in graphs or tables and drawing conclusions) exist in other forms of reporting on inquiry investigations such as Vee-maps (Novak & Gowan, 1984). Therefore, the issues identified with PSTs providing formative feedback for laboratory reports could likely be found with other forms of student reports.

This study provides some further insights into resistance to implementing science inquiry in schools. Having high school students engage in inquiry and write detailed laboratory reports could result in a heavy marking load for teachers. There are strategies to deal with this increased load (e.g., group reports, grading of reports according to a rubric), but there might be a need for these to be explicitly taught to PSTs as a way of dealing with the potential for increased marking load. To help overcome the resistance that many teachers have to inquiry activities (Bybee et al., 2008), PSTs must be given a context (perhaps a PST-specific course on assessment/evaluation) in which they can develop their own understanding of formative assessment as it applies to these inquiry activities in science classrooms. They need to have the opportunity to experience "active learning strategies to engage students in the process of science *and* [emphasis added] teaching methods that have been systematically tested" (Handelsman et al., 2004, p. 521). Without challenging experiences—and the capacity for constructive reflection on those experiences—along with the ability to provide high-quality formative feedback, PSTs' attitudes to engaging their own students in inquiry investigations may be truncated (Bell et al., 2003; Britzman, 1986; Sadler et al., 2010; van Zee & Roberts, 2001).

Acknowledgements The authors wish to thank the science methods students who volunteered to participate in this study. They did so hoping to contribute to the advancement of knowledge and practice in science education, particularly in science teacher education, and we appreciate their faith that together we could do so.

Appendix A: Graphs from Reviewed Laboratory Reports

See Figs. 6.1, 6.2, and 6.3.

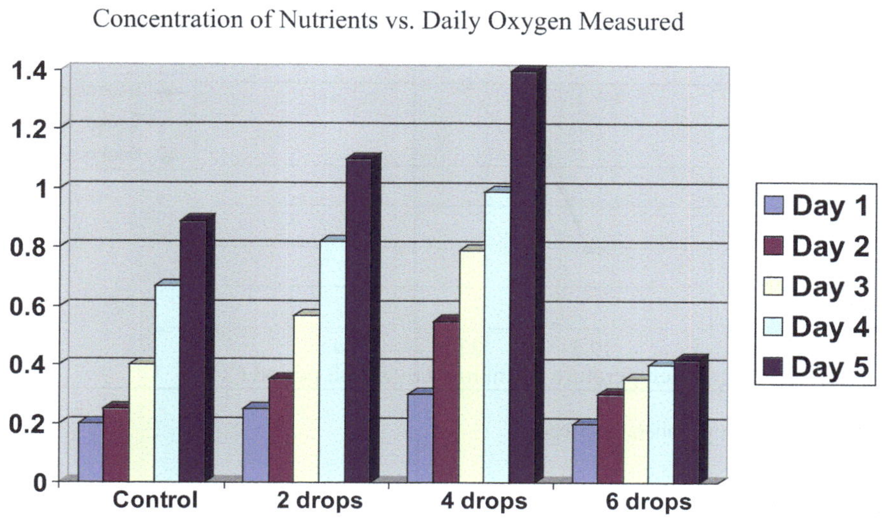

Fig. 6.1 Graph from Laboratory Report A

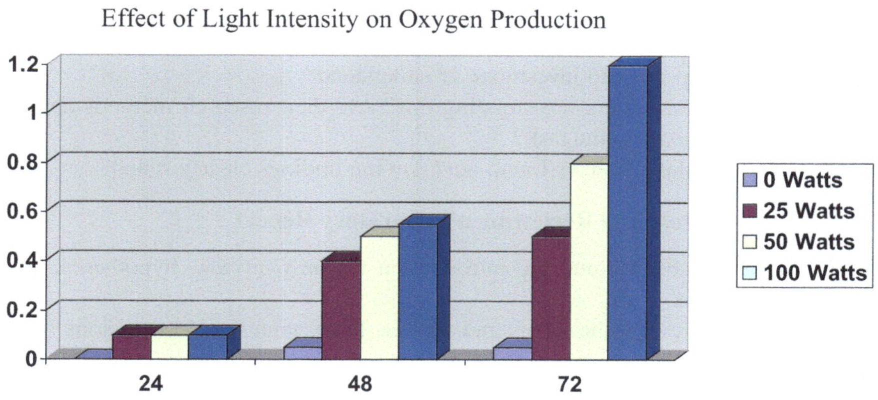

Fig. 6.2 Graph from Laboratory Report B

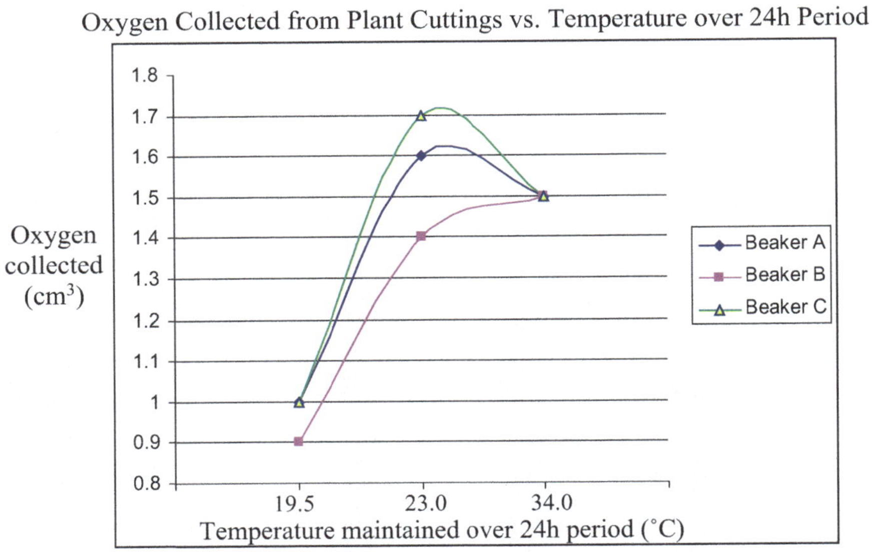

Fig. 6.3 Graph from Laboratory Report C

Appendix B: Suggestions for Reviewing a Laboratory Report

Start by asking *yourself* the following general questions while you read the paper:

1. What do the researchers want to find out?
2. Why is that important to investigate or understand?
3. How are the researchers investigating this? Are their research methods appropriate and adequate to the task?
4. What do they claim to have found out? Are the findings clearly stated?

A. **General Approach to Reviewing a Laboratory Report**

   1. Read the abstract and the introduction for an overview, hypotheses, and conclusions.
   2. Carefully review the tables and figures. Draw your own conclusions from them; don't rely on those of the authors.
   3. Read the methods and results. Make notes on the design. Do written results agree with tables and figures? Do they agree with your conclusions from step 2?
   4. Read the discussion. Do conclusions follow from results? Alternative interpretations? Use of references/theory?

B. **Things to Look for in Each Section**

1. *Title, Abstract, Introduction*

   - What are the questions or hypotheses?
   - Primary value of the study: New science? Technique? Novel idea? Review?
   - Is there an adequate review of previous research or related theory/concepts?

2. *Methods and Results*

   - Are methods clear? Can they be repeated?
   - Are diagrams (needed?) sufficient to explain the research approach used?
   - Kinds of evidence used: Descriptive? Experimental? Both?
   - Sampling and experimental design used: Replication? Controls? Adequate to test hypotheses? Would other designs be better?
   - Numbers: Provided? Used in analyses? Used correctly?
   - Statistics: Used? Appropriate? Correct?
   - Are tables structured appropriately? Are averages or other summary statistics provided (if appropriate)?
   - Is data represented using an appropriate graph (or other summary)?

3. *Discussion (Assumptions and Interpretations)*

   - What assumptions are stated? Implied?
   - Can these be tested? Should they be?
   - Did investigator seek to support hypotheses or attempt to refute them?
   - Is interpretation supported by all the available data?
   - Is there data or a reference for each conclusive statement?
   - Is appropriate "theory" used to support conclusions drawn from the data?
   - Do alternatives exist? Are they considered?
   - Is the author's interpretation of references correct (if you know)?
   - Could the investigator have done better under the circumstances?
   - What future studies are suggested by the outcome of this study?

C. **Structuring Tables**

1. Tables usually show raw data with appropriate summary statistics (averages, counts, frequencies, standard deviation, range, etc.).
2. Usually, but not always, treatments are in individual columns and different replicates/trials are in the rows.
3. Tables are usually structured so that they can be *visually inspected* for patterns in the data *before* the graph is drawn/looked at.
   Examples of table structure:
   For nominal or ordinal studies (usually represented with a bar graph or sometimes a line graph):

	Treatment 1	Treatment 2	Treatment 3	Treatment 4
Rep 1				
Rep 2				
Average:				

For correlational studies (usually represented with a scatterplot sometimes with a trend line):

Variable 1	Variable 2

D. **Choosing an Appropriate Graph**

1. A graph is usually chosen to represent the type of data being presented.
   - It is usually appropriate for a bar graph (and sometimes a pie chart) to be used to represent *nominal* level data (graphs comparing categories such as dogs & cats, males & females).
   - For *ordinal* level data (where there is an unmeasured but changing category of the same thing [such as 'small,' 'medium,' and 'large' rocks where those sizes represent a set range]), it is usually most appropriate to use a bar graph and sometimes a line graph.
   - For *interval/ratio* level data (where both things are measured variables, for instance, arm length in cm and leg length in cm), a scatterplot with a trend line is usually most appropriate.
2. Graphs that have an indicator of the *range of data* in the raw data (either with error bars, or by graphing the raw data and then the averages or lines of best fit [with measures of fit]) are usually better as it allows the reader to understand how strong the patterns of relationship in the findings are. It is important for an author not to overstate or overemphasize the strength of the patterns in the data.

References

Atjonen, P., Pöntinen, S., Kontkanen, S., & Ruotsalainen, P. (2022). In enhancing preservice teachers' assessment literacy: Focus on knowledge base, conceptions of assessment, and teacher learning. *Frontiers in Education, 7*, 891391. https://doi.org/10.3389/feduc.2022.891391

Bell, R. L., Blair, L. M., Crawford, B. A., & Lederman, N. G. (2003). Just do it? Impact of a science apprenticeship program on high school students' understandings of the nature of science and science inquiry. *Journal of Research in Science Teaching, 40*(5), 487–509. https://doi.org/10.1002/tea.10086

Black, P. J., Harrison, C., Lee, C., Marshall, B., & Wiliam, D. (2003). *Assessment for learning: Putting it into practice.* Oxford University Press.

Black, P. J., Harrison, C., Lee, C., Marshall, B., & Wiliam, D. (2004). Working inside the black box: Assessment for learning in the classroom. *Phi Delta Kappan, 86*(1), 8–21. https://doi.org/10.1177/003172170408600105

Black, P. J., & Wiliam, D. (1998a). Assessment and classroom learning. *Assessment in Education: Principles, Policy & Practice, 5*(1), 7–74. https://doi.org/10.1080/0969595980050102

Black, P. J., & Wiliam, D. (1998b). Inside the black box: Raising standards through classroom assessment. *Phi Delta Kappan, 80*(2), 139–148.

Bowen, G. M., & Bartley, A. W. (2007). Understanding issues of pre-service teachers reviewing high school science laboratory reports submitted to an on-line publishing environment. *International Journal of Learning: Annual Review, 14*(9), 45–58. https://doi.org/10.18848/1447-9494/CGP/v14i09/45478

Bowen, M., & Bartley, A. (2013). *The basics of data literacy: Helping your students (and you!) make sense of data.* National Science Teachers Association.

Bowen, G. M., Bartley, A. W., MacDonald, L., & Sherman, A. (2016). Experiences with activities developing pre-service science teacher data literacy. In G. A. Buck & V. L. Akerson (Eds.), *Enhancing professional knowledge of pre-service science teacher education by self-study research: Turning a critical eye on our practice* (pp. 243–269). Springer. https://doi.org/10.1007/978-3-319-32447-0_13

Bowen, G. M., & Roth, W.-M. (2005). Data and graph interpretation practices among pre-service science teachers. *Journal of Research in Science Teaching, 42*(10), 1063–1088. https://doi.org/10.1002/tea.20086

Bradley, D. (2005). *Practicals in science education: A study of the theoretical bases, rationale and implementation of practicals in junior secondary science education* [Unpublished doctoral dissertation]. Curtin University of Technology.

Britzman, D. P. (1986). Cultural myths in the making of a teacher: Biography and social structure in teacher education. *Harvard Educational Review, 56*(4), 442–457. https://doi.org/10.17763/haer.56.4.mv28227614l44u66

Bybee, R. W., & Carlson Powell, J., & Trowbridge, L. W. (2008). *Teaching secondary school science: Strategies for developing scientific literacy* (9th ed.). Pearson.

Çil, E., & Kar, H. (2015). Pre-service science teachers' interpretations of graphs: A cross-sectional study. *Science Educator, 24*(1), 36–44. http://files.eric.ed.gov/fulltext/EJ1069988.pdf

Dassa, C., Vázquez-Abad, J., & Ajar, D. (1993). Formative assessment in a classroom setting: From practice to computer innovations. *Alberta Journal of Educational Research, 39*(1), 111–125.

Daws, N., & Singh, B. (1996). Formative assessment: To what extent is its potential to enhance pupils' science being realized? *School Science Review, 77*(281), 93–100.

Deters, K. M. (2005). Student opinions regarding inquiry-based labs. *Journal of Chemical Education, 82*(8), 1178–1180. https://doi.org/10.1021/ed082p1178

Espiritu, M. J. B., Shahrill, M., Perera, J. S. H. Q., & Prahmana, R. C. I. (2018). Formative assessment in science education: Is it being practiced? *Journal of Physics: Conference Series, 1088*(1), 012009. https://doi.org/10.1088/1742-6596/1088/1/012009

Gericke, N., Högström, P., & Wallin, J. (2023). A systematic review of research on laboratory work in secondary school. *Studies in Science Education, 59*(2), 245–285. https://doi.org/10.1080/03057267.2022.2090125

Handelsman, J., Ebert-May, D., Beichner, R., Bruns, P., Chang, A., DeHaan, R., Gentile, J., Lauffer, S., Stewart, J., Tilghman, S. M., & Wood, W. B. (2004). Scientific teaching. *Science, 304*(5670), 521–522.

Hofstein, A., & Lunetta, V. N. (2004). The laboratory in science education: Foundations for the twenty-first century. *Science Education, 88*(1), 28–54. https://doi.org/10.1002/sce.10106

Jones, D., Kaplanis, C., Melville, W., & Bartley, A. W. (2009). Science as inquiry at Sir Winston Churchill Collegiate and Vocational Institute. In R. E. Yager (Ed.), *Inquiry: The key to exemplary science* (pp. 151–176). National Science Teachers Association.

Jordan, B., & Henderson, A. (1995). Interaction analysis: Foundations and practice. *Journal of the Learning Sciences, 4*(1), 39–103. https://doi.org/10.1207/s15327809jls0401_2

Luft, J. A. (2001). Changing inquiry practices and beliefs: The impact of an inquiry-based professional development programme on beginning and experienced secondary science teachers. *International Journal of Science Education, 23*(5), 517–534. https://doi.org/10.1080/09500690121307

Lunetta, V. N. (1998). The school science laboratory: Historical perspectives and contexts for contemporary teaching. In K. Tobin & B. Fraser (Eds.), *International handbook of science education* (pp. 249–264). Kluwer/Springer.

Máñez, I., Lipnevich, A. A., Lopera-Oquendo, C., & Cerdán, R. (2024). Examining pre-service teachers' feedback on low-and high-quality written assignments. *Educational Assessment, Evaluation & Accountability, 36*(2), 225–256. https://doi.org/10.1007/s11092-024-09432-x

Melear, C. T. (1999). Middle and secondary science teacher opinions about working in scientists' laboratories during preservice preparation. *The Electronic Journal for Research in Science & Mathematics Education, 4*(2). https://ejrsme.icrsme.com/article/view/7627

Melville, W., Jones, D., & Campbell, T. (2014). Distributed leadership with the aim of 'reculturing': A departmental case study. *School Leadership & Management, 34*(3), 237–254. https://doi.org/10.1080/13632434.2013.849681

National Research Council. (2013). *Next generation science standards: For states, by states.* National Academies Press. https://doi.org/10.17226/18290

National Research Council. (2014). *Developing assessments for the next generation science standards.* National Academies Press. https://doi.org/10.17226/18409

Novak, J. D., & Gowin, D. B. (1984). Learning how to learn. *Cambridge University Press.* https://doi.org/10.1017/CBO9781139173469

Parkinson, J. (2017). The student laboratory report genre: A genre analysis. *English for Specific Purposes, 45*, 1–13. https://doi.org/10.1016/j.esp.2016.08.001

Ruelmann, M., Charalambous, C. Y., & Praetorius, A. K. (2023). The representation of feedback literature in classroom observation frameworks: An exploratory study. *Educational Assessment, Evaluation and Accountability, 35*(1), 67–104. https://doi.org/10.1007/s11092-022-09403-0

Roth, W.-M., McGinn, M. K., & Bowen, G. M. (1998). How prepared are preservice teachers to teach scientific inquiry? Levels of performance in scientific representation practices. *Journal of Science Teacher Education, 9*(1), 25–48. https://doi.org/10.1023/A:1009465505918

Sadler, T. D., Burgin, S., McKinney, L., & Ponjuan, L. (2010). Learning science through research apprenticeships: A critical review of the literature. *Journal of Research in Science Teaching, 47*(3), 235–256. https://doi.org/10.1002/tea.20326

Shavelson, R. J., Young, D. B., Ayala, C. C., Brandon, P. R., Furtak, E. M., Ruiz-Primo, M. A., Tomita, M. K., & Yin, Y. (2008). On the impact of curriculum-embedded formative assessment on learning: A collaboration between curriculum and assessment developers. *Applied Measurement in Education, 21*(4), 295–314. https://doi.org/10.1080/08957340802347647

Staberg, R. L., Febri, M. I. M., Gjøvik, Ø., Sikko, S. A., & Pepin, B. (2023). Science teachers' interactions with resources for formative assessment purposes. *Educational Assessment, Evaluation & Accountability, 35*(1), 5–35. https://doi.org/10.1007/s11092-024-09432-x

Strauss, A. L., & Corbin, J. M. (1990). *Basics of qualitative research: Grounded theory procedures and techniques.* Sage.

van Zee, E. H., & Roberts, D. (2001). Using pedagogical inquiries as a basis for learning to teach: Prospective teachers' reflections upon positive science learning experiences. *Science Education, 85*(6), 733–757. https://doi.org/10.1002/sce.1036

Volkmann, M. J., & Abell, S. K. (2003). Rethinking laboratories: Tools for converting cookbook labs into inquiry. *The Science Teacher, 70*(6), 38–41. http://www.jstor.org/stable/24156087

Wilcox, B. R., & Lewandowski, H. J. (2016). Open-ended versus guided laboratory activities: Impact on students' beliefs about experimental physics. *Physical Review Physics Education Research, 12*(2), 020132. https://doi.org/10.1103/PhysRevPhysEducRes.12.020132

Woolnough, B. E., & Allsop, T. (1985). *Practical work in science.* Cambridge University Press.

Zemel, Y., Shwartz, G., & Avargil, S. (2021). Preservice teachers' enactment of formative assessment using rubrics in the inquiry-based chemistry laboratory. *Chemistry Education Research & Practice, 22*(4), 1074–1092. https://doi.org/10.1039/D1RP00001B

G. Michael Bowen, BSc (University of Guelph), BJ (University of King's College), BEd (Dalhousie University), MSc (University of Guelph), MA (University of Guelph), PhD (University of Victoria) is an Associate Professor at Mount Saint Vincent University in Halifax, Nova Scotia. He teaches in the Bachelor of Education program (including courses in secondary science methods, science content, STEM/technology, and media literacy) and in the graduate program (including courses in curriculum studies, research methodology, and identity theory). In addition, he has worked in informal science education (working in the whale watching industry in Atlantic Canada) and has also taught science from Grades 5 to 13 in Canada and the United States. His research areas include science teacher education, learning science in authentic contexts (from elementary to post-doctoral levels), use of STEM technologies in early learners, informal science education, graph/data literacy, (news)media portrayals of science, and citizen science.

Anthony W. Bartley, BA (University of Essex, UK), PGCE (University of Liverpool, UK), MSc (University of Kent, UK), PhD (University of British Columbia) is an Emeritus Professor of Education at Lakehead University. He has over 16 years of classroom experience in the UK and Canada, before doctoral studies and a faculty position at Lakehead in August 1994. His research interests have included alternative assessment, preservice teacher education focusing on teaching science through inquiry, and data literacy. He was a member of The Ontario Education Research Panel from 2013 to 2015. Retirement came in December 2017; he is now messing about in boats on the Pacific Coast.

Wayne Melville, BSc (University of Tasmania), DipEd (University of Tasmania), Grad Cert Ed Admin (University of Adelaide), MSc (Curtin University of Technology), PhD (Curtin University of Technology), has been the Dean of the Faculty of Education at Lakehead University in Ontario since 2019. He taught science in Australian secondary schools from 1989 to 2005. In 2005 he commenced his academic career at Lakehead, working in secondary science initial teacher education. He was granted Lakehead University's Distinguished Instructor Award in 2017 and has published over 65 refereed articles and four books. He also served as the Co-Editor in Chief of the Journal of Science Teacher Education from 2019–2024.

Chapter 7
News Media and Its Depiction of Science: A Case in a Canadian Secondary Biology Classroom

Giuliano Reis, Pei-Ling Hsu, and Wolff-Michael Roth

7.1 Introduction

News media, whether print or online, can be considered valuable resources for science education in that they can connect the science curriculum to everyday life. However, little is known about how science teachers' use of news media in the classroom can mediate their students' understanding of the nature of science (NOS) and the social implications of the scientific endeavour. Using a Canadian secondary biology teacher as an exemplary case, this chapter describes and examines how available discursive elements—both textual and visual—of a news report about a scientific discovery can be used to make sense of the NOS and the influence of science on society. It is an exploratory case study, a kind of qualitative research approach, where we embrace discourse analysis as our theory and method of choice by relying on video-recorded classroom observations, news media reports, and participants' talk as the main sources of data. Our discussion focuses on the potential influence that teachers' recounts and reinterpretations of news media stories can have on their students' reconstruction of a particular view of the NOS as well as on their perceptions of the purpose of science in our societies. Moreover, we illustrate how discourse is more than words because it is irreducibly intertwined with the human activity to which it belongs. In showcasing how science education is realized within/through people's

G. Reis (✉)
University of Ottawa, Ottawa, ON, Canada
e-mail: greis@uottawa.ca

P.-L. Hsu
University of Texas - El Paso, El Paso, TX, USA
e-mail: phsu3@utep.edu

W.-M. Roth
University of Victoria, Victoria, BC, Canada
e-mail: mroth@uvic.ca

C. D. Tippett and T. M. Milford (eds.), *Seeing Science Through the Eyes of Canadian Teachers and Learners*, Contemporary Trends and Issues in Science Education 54,
https://doi.org/10.1007/978-3-031-88785-7_7

day-to-day interactions, our chapter exemplifies how discourse analysis is a valuable theoretical and methodological tool for science education research.

7.1.1 A Media–ted Society

Currently, more than half of Canadians follow the news daily (Statistics Canada, 2023), with adults (16–64 years old) spending an average of 1 h and 15 min per day reading online and physical print news (Guttmann, 2024). News Media Canada (2022) reported the existence of over 1,000 newspaper titles in the country with a weekly circulation totaling almost 30 million copies. Hence, news media consumption is a common activity and one by which the public can learn about the latest scientific advancements and the most pressing scientific issues (Shea, 2015).

Notably, science education is not impervious to the pervasiveness of the Canadian news media landscape. For example, the Ontario science curriculum document for Grades 1–8 states:

> Opportunities should be provided for students to engage in a critical discussion of *texts* [emphasis added], including books and textbooks, television programs, movies, documentaries, web pages, advertising, music, gestures, oral texts, newspaper and magazine articles, letters, cultural text forms, stories, and other forms of expression. Such discussions empower students to understand the impact on members of society that was intended by the text's creators. Language and communication are never neutral: they are used to inform, entertain, persuade, and manipulate. (Ontario Ministry of Education [OME], 2022b, p. 32).

The underlying message in the Grade 9 science curriculum document is no different:

> Considering the vast and ever-increasing number of sources of information available today, students need to be aware of how to find and identify appropriate information during research. Critical-thinking skills are essential to assess the information gathered, in part by considering the biases, interests, and motivation of the authors, as well as the trustworthiness of the source or publisher. Students should also carefully consider how scientific knowledge is shared, whether in formal, peer-reviewed contexts or through less formal channels such as social media. (OME, 2022a, p. 22).

Clearly, media (including news media) expected to play a role in science education for Ontario students in Grades 1–9 as informal avenues through which scientific knowledge is communicated to the public.

As these science curriculum documents suggest, caution about science-related reports circulated in the media, including news media and social media, is necessary because those reports can be inaccurate. For example, the combination of the increasing influence of social media with the spreading of fake news about the COVID-19 pandemic proved to be problematic from a science and science education standpoint (Otte & Beeken, 2022). Public perception of the trustworthiness of science was negatively impacted while scientists worldwide raced to development a vaccine.

As for media producers, they use effective strategies to sell their stories—or sell audiences—to advertisers (Fairclough, 1995), thereby creating a *media-ted* reality that can be indistinguishable from real life (Lind, 1998). This, in turn, makes news media a likely powerful influence that can be both an indicator of public thinking about science and a source for that thinking. Even though science news media is often assumed to present scientific information in a factual manner, it may also be intended to persuade the audience to reach desired conclusions (van Dijk, 1991). Persuasion is a key element in the understanding that news media is part of a larger process of marketing with ambiguous information when the opposing interests of owners and audiences arise from a competitive struggle for reader attention. As a result, any gaps in news media reports may be filled by readers who use their knowledge and experience to assign meaning (Gee, 2001).

Thus, news media content reflects its producers' conceptions of what audiences want or need to know about science and reveals what those audiences are expected to understand about science (e.g., how scientists work, the role of science in society, etc.). In elevating or downplaying scientific information in certain ways, news media affects the way that people interpret the importance and accuracy of that information as well as their inclination to seek sources of information that align with their particular interpretations (e.g., Einsiedel, 1992; Jarman & McLune, 2007).

Science and science education are not dissociated from news media (e.g., Mackenzie et al., 2020); that is, news media is an important source of information about science-related issues (Dempster et al., 2022; Wellington, 1993). Likewise, news media can contribute to advancing citizens' science literacy by influencing public understanding of science and its nature and function in society (e.g., Demirdöğen & Aydın-Günbatar, 2021). The potential impact of news media on science literacy warrants research on its use in science education.

7.1.2 A Media-Ted Science Education?

Even though news media is not designed as an educational tool per se (Wallace, 2004), it can support science teachers' efforts to help students make sense of science content by linking the science curriculum with current events (e.g., Elliott, 2006); that is, news content can be relevant to science teaching and learning (e.g., Arnold, 2019). News media can provide a means for linking scientific issues to students' prior experiences and has the potential to create interest in science (e.g., Retzbach & Maier, 2015). Notably, the use of science news media in the classroom to mediate understanding of the NOS and the role of science in society has received little research attention.

In the context of science education, the interpretation of news media reports is often mediated by teachers (Ratcliffe, 1999), who can support students by drawing on events in and out of the classroom or by considering additional information to understand the news. In analyzing the ways that news media reports are used in science classrooms, we can learn what the existing gaps are between the contents of news media reports and what teachers and students already know and consider the potential implications of news media use for understanding the NOS and the many implications of science in society.

In this chapter, we turn our gaze to the "everyday, practical, concrete, and seen-but-unnoticed interactional work [discursive practice] of making sense of mass media" (Cousineau, 2015, p. 36) in order to contribute to conversations about the educational use of science news media (e.g., Kachan et al., 2006). More specifically, we look at the case of Elizabeth (pseudonym), a Grade 11 Biology teacher, who uses a science-related newspaper report in her classroom. We describe and analyze the varied textual and visual discursive elements of the news report that Elizabeth draws on to make the content more accessible to her students. In addition, we consider how Elizabeth's recount and reinterpretation of the news media report can serve to mediate her students' views of the NOS and their perceptions of the social importance of scientific activities.

7.1.3 Discourse Analysis as a Language-Based Research Method

We conducted a discourse analysis—a method that investigates how meaning is constructed through language and "explores the functions, operations, and powers of language in various contexts, examining its impact on social relationships and power structures." (Macrae, 2020, para. 1). According to Lester and Paulsen (2018), discourse analysis assumes that:

> language is performative; that is, it is assumed to always *do* something. For instance, through language people go about ascribing blame, building accounts, negotiating peace, and so on. Relatedly, discourse analysts generally assume [a social constructionist position] that it is through language that the social world is built, ordered, sustained, reframed, and so on. (p. 58).

Like Berger and Luckmann (1967), we believe that talk and text are used to produce the social world. Therefore, by focusing on what is said and written as well as on how it is said and written, we can articulate the functions of discourse within the context in which its embedded (Potter, 2003).

7.2 Methods

Our exploration of the use of a news media report in a science classroom provides an example of the ways in which discourse analysis can used in science education. We asked the following questions:

1. What textual, pictographic, and organizational resources were available for Elizabeth to make sense of what she read?

2. Which of those resources did she choose to draw upon during her reading?
3. How does what she read and the way that she read it contribute to creating a particular representation of science, its nature, and its importance in society to her students?

The reading that Elizabeth did represents an authentic, naturalistic situation in that she performed it as part of her teaching practice and not exclusively for this research. What we observed was what Elizabeth considered relevant at the time to facilitate her students' science learning. The news media report that she shared with her class, about dogs that can diagnose the presence of certain types of cancer in humans, is particularly relevant for several reasons. First, health-related themes are amongst the most common in the news media used in classrooms (Jarman & McClune, 2002). Second, cancer-related issues draw media attention because of cancer's prevalence: an estimated 247,000 new cases of cancer are projected to occur in Canada in 2024 (Canadian Cancer Society, 2024). Third, dogs are animals with which people generally have strong relationships (Monteiro & Reis, 2018).

At the time of data collection, Elizabeth was teaching at a public school in a midsized Canadian city. She had nearly three decades of teaching experience and was the head of the school's science department. In her Grade 11 Biology class, which met daily for 90 min during the semester, there were 28 students; 20 of whom were in a biology career preparation program that was intended to provide enriched, career-specific learning opportunities. Based on Elizabeth's classroom composition, it is safe to infer that her teaching was aimed at developing the skills and knowledge necessary for students to enter the scientific workforce or pursue further education in scientifically oriented fields.

As part of a larger research project on secondary students' internships with scientists, Elizabeth was interviewed by the second author (Roth & Hsu, 2010). During the interview, she accounted for her use of news media, specifically newspapers, in her science classroom:

> I used to like to bring in newspaper articles all of the time, to relate science to everyday life, and then to talk also about the knowledge aspect of science, and the process aspect of science, and how that informs, you know, so much, so much of their environment, and their lifestyle, and … but also, always sort of with the disclaimer that it's one way of looking at the world.

Here, Elizabeth acknowledges the advantages of using newspaper articles to "relate science to everyday life." That way, she indicates the existence of a possible gap between what students learn in her class and "their lifestyle" – one that is filled by discussing "the knowledge … and the process aspect[s] of science." It appears as if Elizabeth is trying to make her teaching and her students' learning relevant. Without elaborating further, she points to another distinction within her discipline: students must understand that there is a "knowledge" and a "process" aspect to science. In the end, she is clear that science is but one way of "looking at the world." Rather unfortunately, she does not specify any other ways of looking at the world that she discusses in her classes: perhaps religion (e.g., Reis et al., 2023) or Indigenous Traditional Knowledge (e.g., Reis & Ng-A-Fook, 2010).

Fig. 7.1 Artistic rendition of Elizabeth in front of her class

We also observed Elizabeth during a science class. After entering the room, she moved to the front of her desk. She holds up a small piece of paper by her index finger and thumb, with her palm partly open, as if wanting to make it visible to all. The other hand holds a larger piece of paper as seen in Fig. 7.1.

Elizabeth addresses the class[1]:

> Alright. (3s) The news. Front page. ((Shows students a small cut out from a newspaper)) Did anybody notice yesterday's front page? 'The nose knows' with the puppy here, the dog? So (3s) ((Reads from a bigger piece of newspaper)) 'Dogs show ability to sniff out cancer.' So, remember our presentation before Christmas on the sniffer dog? (3s) ((Looks around)) So, if I could just explain a little bit about this. (2s) This links back to the work that we did with a presentation on the ability of dogs to have this phenomenal um response, right? Amazing numbers of receptors and structuring their noses to allow them to detect. It says here: 'Dogs can detect some …' There's only one person talking right now, it's me. 'Dogs can detect some chemicals diluted to a few parts per billion.' So, like little individual molecules basically that, you know, they're detecting, and they actually use the same strategy that you saw in this news for detecting the presence of drugs, ((looks at the paper she is holding)) the dogs were trained, um they don't have to come up and sniff the person's breath, the person breathes into a Ziplock bag and then the dogs test the air in the Ziplock bags and when they detect certain, (4s) ((clears throat)) (2s) when they detect certain chemicals that are given off by the body, right?, when the cancer cells are present, um they sit down. And if they don't detect those chemicals, they move on to the next test bag, ok? You know, so same strategy.
>
> Um … so, another um point here then it ties into biology is that um we've been looking at respiratory systems and talking about how animals need to get rid of basically nitrogenous waste that are breakdown products of when we digest protein, but there are other chemicals, right?, that are given off as your body is just carrying on its biochemical business and if you have cancer, then there are certain specific um chemical substances that um the body has to get rid of as waste. These include benzene and some other organic compounds, and that's what the dogs that are trained to detect and it they came up with this idea because stories were around about people who had skin cancer and have a dog, and the dog would just be

[1] Transcription symbols: () = represents short pauses, where the number inside indicates the length in seconds; (()) description of actions.

very interested in the smell of this tumour on the skin and then once the tumour was removed they have their surgery, the dogs lost interest in it again.

So, that kinda led some researchers to, you know, to actually test this out and run a controlled experimentExperiments. It's a very small experiment, only five dogs were used um however um you know it's also that's other information, there's um the Italian group that were, they're working on trying to come up with a machine, a breathalyser kind of machine, to analyze for these specific compounds that are known to be given off, right, by the body when cancerCancer is present. The dogs were 90% accurate, which is very good, a lot of um tests that you have for things are not that accurate, right? So, so that was considered. They used labradors and they used Portuguese water dogs, so they figured probably most dogs would be able to do it. Of course, biologists have to be funny, so they've two nicknames for this: dognosis and lab testing. ((Some students laugh as the teacher turns away to her desk)). (3 s) So (3 s) um (4 s) Okay, I think that's only introductory stuff.

Overall, this initial sequence of actions by the teacher has three distinct moments:

1. introduction of the science news report by largely situating it within her course ("remember our presentation before Christmas on the sniffer dog? […] This links back to the work that we did with a presentation on the ability of dogs to have this phenomenal um response, right?")
2. discussion of the "knowledge aspect of science" (e.g., biochemistry of the respiratory system)
3. reference to the "process aspect of science" (e.g., source of the idea for the experiment with dogs, the replicability of the tests, and the accuracy and possible generalizability of the results obtained).

7.3 Analysis and Discussion

In this section, we take a closer at how Elizabeth's talk while reading a science news report to her students illustrates the use of available cultural resources afforded by this type of media. Which parts did she change? Which parts did she leave out? How does her reading contribute to creating a certain understanding of the NOS by her students? What sort of role does her reading ascribe to science in society?

Elizabeth's read-aloud seems to conform with what she claimed during her interview: the news report she chooses to share with her students in class allows her to create a link between the science (biology) curriculum and everyday life; in this case, how scientists are working to create more efficient and cost-effective ways to early detect cancer. In addition, her short monologue addresses the two inherent but distinguishable aspects of science: knowledge and process.

When saying "The news. Front page," there are no connectors between these two sentences, which turns Elizabeth's opening lines into fragments. Much like the headlines of a newspaper, she draws students' attention to what she holds in her right hand by economically describing the teaser shown in Fig. 7.2. The fragmentary phrases foretell that something is coming and, thus, constitute a resource for creating anticipation that is a context for attention. Indeed, the sentence fragments make immediate sense when hearers' experiences include newspapers that, unlike other

The nose knows

Here's one more reason to love our four-legged friends. In a study conducted by California scientists, dogs showed an uncanny ability to detect cancer, even in its early stages, by smelling the breath of patients for telltale chemicals given off by cancer cells.

Story, A2

Dogs show ability to sniff out cancer

BY TOM SPEARS
CanWest News Service

OTTAWA — Dogs can often detect cancer, even in its early stages, by smelling the breath of patients for telltale chemicals given off by cancer cells, a clinical trial in California shows.

The Portuguese water dogs and Labradors could, after brief training, sniff out which people had lung or breast cancer with accuracy rates between 88 and 99 per cent — better than some laboratory tests. And they weren't thrown off by cigarette odour in breath.

Now the California scientists who did the study suggest this sniffing — nicknamed dognosis, as well as Lab testing — could be an extra screening tool for detecting cancer's first stages.

All cells give off waste chemicals. But cancer cells give off some distinct ones — a mix of organic vapours such as benzene (also found in gasoline) and alkenes, which the body flushes away by exhaling.

Dogs can detect some chemicals diluted to a few parts per billion in air. Scientists have heard anecdotal reports for years about dogs that curiously sniffed owners who developed skin cancer, and stopped when the tumours were removed.

In Italy, researchers have been experimenting for at least three years with a machine to analyse the breath as a way to diagnose cancers.

Now five dogs tested by the Pine Street Foundation, a California cancer research group, showed an ability to identify or rule out lung and breast cancer correctly, at both early and late stages, about 90 per cent of the time. Both dog breeds performed at the same level.

And the scientists say their success with early-stage cancer is key, meaning the test could be valuable for detecting cancer when treatment has its best chance for success.

"Training was efficient and cancer identification was accurate," the team writes. "In a matter of weeks, ordinary household dogs with only basic behavioural 'puppy training' were trained to accurately distinguish breath samples of lung and breast cancer patients from those of controls (healthy people)."

In lung cancer, the study notes, chest X-rays and sputum tests (analysing mucus from the lungs) "have a high false-negative rate and therefore fail to detect many early-stage cases." Mammograms for breast cancer aren't perfect either, it says.

The handlers made five dogs (three Labs, two Portuguese water dogs) sniff the breath of 55 people with lung cancer, 33 with breast cancer, and 83 healthy people. The breath samples were collected in Ziploc bags; dogs and study volunteers never met. In all, the dogs sniffed samples more than 12,000 times over four months. They learned to sit or lie down if they smelled cancer, and to move on to the next sample if they didn't.

Fig. 7.2 Items from the Times Colonist newspaper (January 12, 2006), teaser (left) and news media report (right) regarding cancer sniffing dogs

types of news media (e.g., radio), have a front page. Those fragments orient hearing to what comes next in a particular way, again, as something having to do with the latest news appearing on the front pages of newspapers (Roth, 2010).

The way Elizabeth holds up the clipping of the news media report suggests that she wants her students to notice it. This gesture, in turn, reinforces the teacher's call for attention. When asking if "anybody [had noticed] yesterday's front page," Elizabeth does not mention the publication's name, leading one to believe that it is either a newspaper title that she commonly brings to class or one to which she knows her students have access. Either way, she has an expectation that one of her students would have noticed it; in which case, they would have been expected to associate what they had learned in school with the information in the news media report. The collection of Elizabeth's initial actions, specifically, the way she holds the teaser and how she refers to it, signals that the object in her hand is worthy of the students' consideration. After all, only important (i.e., relevant) news makes the front page of newspapers.

Elizabeth proceeds to read the headline of the teaser ("the nose knows") and refers to the picture that accompanies it ("the puppy here, the dog"). Therefore, both picture and text function as references (i.e., intertexts) to one another; they are being used as backgrounds against which they can be read and understood (Roth et al., 2005). That is, the picture (i.e., a dog's face in close-up) can help students to infer whose nose the headline talks about while directing attention to the news media report. News media reports without pictures tend to be disfavoured by readers (e.g., Halkia & Mantzouridis, 2005).

The teaser headline is an affirmative sentence leaving little or no room for speculation: the nose *does* know. In fact, the headline uses an idiomatic expression that refers to a situation where speakers do indeed know something that the designated recipients did not think they knew. However, the sentence is also uncertain; what exactly does the nose know? This missing piece of information can potentially persuade interested students to continue listening to Elizabeth with the possibly of learning more details.

Elizabeth continues to the headline from the full news media report on the next page of the newspaper: "Dogs show ability to sniff out cancer." At this point, she connects the news media report with a presentation that took place in her classroom a few weeks previously that involving a "[drug] sniffer dog." This reminder, in turn, further legitimizes her decision to share the news with her students because it represents an extension of an activity that was completed as part of the Biology 11 curriculum. Notably, the use of the extreme adjectives "phenomenal" and "amazing" that were not included in the original report to describe the dogs' aptitude to detect cancer and drugs; this could be a strategy to make the media report more interesting.

As a secondary science teacher, it is not surprising that Elizabeth uses specialized terminology during her discourse about how the news media report ties into biology; she talks about receptors, molecules, the respiratory system, nitrogenous waste, protein, benzene, and organic compounds. Likewise, she speaks of the way that scientists worked in the case of cancer-sniffing dogs: from how "they came up with this idea" of studying the early detection of cancer by dogs to how they conducted a "controlled … very small experiment." There was also a group of Italian researchers that was attempting to develop a cancer-detecting machine based on the same principles. That a scientific experiment, however "small," is the grounds for assigning accuracy to the information presented.

Noticeably, Elizabeth's reading subtly reinforces the distinction between an "experiment" and those "stories … that kinda led" to actual tests. Her talk suggests that only scientists are in the privileged position to verify the reliability of such anecdotes and that everything else is circumstantial. This emphasis, in turn, confers authority to the story and stresses the image of scientists and the work they do as important (i.e., relevant) for society (Conrad, 1999). The mention of two different teams of researchers working on similar early cancer detection projects points to the relevance of a number of the attributes of doing science: collectivity (i.e., scientists don't work alone), creativity (i.e., there is more than one way to research a problem), replicability (i.e., tests need to be carried out multiple times to determine the accuracy of results), and generalizability (i.e., the potential that other domesticated animals may be able to detect cancer just the same). The use of the adverb "however" before the information about the Italian group in combination with the statistic that dogs were 90% accurate—considered by the teacher as "very good"—likely serves to make up for the "very small" scale of the experiment. The potential generalizability of the results—scientists "figured probably most dogs would be able to do it"—are used to strengthen the relevance of the study and of the science news media report for the class.

Elizabeth finishes her reading by stating that certain scientists have an obvious sense of humour: "of course, biologists have to be funny." She considers the play with words (i.e., "dognosis and lab testing") amusing; although she does not explicitly explain the humour, students are heard laughing at the wordplay. Humour, a powerful science communication tool (Meyer, 2000), helps Elizabeth to share information not mentioned in the news report (i.e., scientists have a comical side).

It appears safe to assume that the elements in Elizabeth's discourse are presupposed to be understandable by her students as well as linkable to their reality both in and out of school. The comparison between what and how she reads the news media report points to what she may consider its important elements. Conversely, the comparison also provides a sense of the information she considers to be of less value for her students.

Examining the original news media report (Fig. 7.2), it is clear that Elizabeth reinterpreted, modified, and omitted some of the information that she read to her students, thus mediating their understanding of the NOS and the role of science in society. For instance, certain adjectives she uses are not present in the actual publication, nor are other words like "molecules," "nitrogenous" or "protein." Additionally, the original news media report does not mention what type of scientists (e.g., biologists, medical doctors, or veterinarians) participated in the studies mentioned.

Our analysis of both the included and excluded aspects of Elizabeth's discourse suggests that she places importance on both the knowledge and procedural aspects of science (e.g., Bell, 2009). She seeks to connect the relevance of science to her students' lives and their potential future scientific careers. Moreover, her discourse approximates scientists and what they do to students' reality in two ways: by indicating that their research can fall in the domain of biology, the course's topic, and by proposing that scientists are creative and have a sense of humour, both desirable human traits. Elizabeth's discourse is showcasing an example of what students might experience should they choose a career in biology-related fields.

Elizabeth's overall depiction of the NOS seems to follow that of the news media report: science is a collective endeavour that draws on different research methods (e.g., designing experiments based on anecdotal evidence), is highly accurate (90% in this case), replicable, and generalizable. Elizabeth's decision to start her class by pointing out the teaser headline, which contained a conclusive statement about the study, helped to set this tone.

No criticism of the study or the way it was represented in the news media report was presented. For instance, although the sniffing protocol was repeated thousands of times, neither Elizabeth nor the news media report mention the possible reasons for using a small sample when scientific experiments often involve much larger samples. Likewise, not enough information is given about the Italian research to allow meaningful comparisons between the two teams' results. Another aspect of the experiments that was not touched upon was the difference between generalizability and conclusiveness. One of many misconceptions related to the NOS is that findings that are generalizable are also conclusive (McComas, 1996).

Elizabeth's discourse almost imperceptibly hints at the aspects of uncertainty and tentativeness of the scientific endeavour. First, she uses words (i.e., few, like

… basically, kinda) that may reflect the inexactitude of science. Second, she labels the Italian team's efforts to develop a breathalyzer as a work in progress ("they are working on trying"). Perhaps unbeknownst to Elizabeth—there was no way for us to ascertain—her allusion to the uncertainty (i.e., tentativeness) inherent in science is considered a desirable element of science communication (e.g., Retzbach & Maier, 2015). It is not uncommon for news coverage to omit information about uncertainties and to present scientific findings as being beyond doubt (e.g., Schäfer, 2017). Third, in the interview, Elizabeth recognized that science is "one way to look at the world," acknowledging the possibility of other perspectives.

The way in which Elizabeth reads the news media report, omitting or adding information, reveals the importance she places on certain aspects. These differences may be her efforts to link her reading with her course by recalling a previous activity and using specialized vocabulary. These choices are how Elizabeth contextualizes science in the news media report for her students.

7.4 Concluding Thoughts

Our purpose in this chapter was to describe and examine how available textual and discursive elements of a science news media report were used to inform student understanding of the nature of science and how science affects society (and vise-versa). Our analysis revealed that Elizabeth's classroom discourse painted a particular picture of scientists and their work. Accordingly, science, which is a unique way of looking at the world (Leite et al., 2024), is based on the collaborative work of scientists who may be in geographically diverse settings to perform accurate, creative, replicable, and generalizable observations and experiments of real-world phenomena. At the same time, Elizabeth points to scientists as being funny and tentatively acknowledges that science is uncertain, while portraying it as accurate and replicable.

As Elizabeth noted in her interview, news media can play an important role in science education as a source of information and a connection between school and the real world. As our analysis showed, there is room for news media's portrayal of the NOS to be more explicitly addressed and dissected in the classroom. For example, teachers might facilitate discussion about how science and the work of scientists are described or not described. Another teaching opportunity arises in the examination of the ways in which news media can influence the construction of readers' worldviews (Luke, 1996) by using particular strategies to help readers make sense of information. Teachers could show students how to identify the strategies employed by news media outlets to influence the interpretation of their reports (e.g., how pictures and headlines are created and work together to draw one's attention to the news media report).

Like journalists, who describe science based on their judgement on what will make their stories most appealing to readers, Elizabeth too has to decide how to connect science news media to her students' experiences. We do not mean to imply that Elizabeth should have managed her retelling of the news story differently; we simply offer one interpretation of her reading of a news media report in her science classroom through the lens of discourse analysis. Readers, including science education researchers, science teachers, science teacher educators, and preservice science teachers, can use our findings to inform their own use of news media reports to improve science literacy, teaching about the nature of science and the social implications of the scientific endeavour.

Acknowledgements This study was supported by grants from the Natural Sciences and Engineering Research Council of Canada and the Social Sciences and Humanities Research Council of Canada.

References

Arnold, R. (2019). *Fake news in science and education: Leaving weak thinking behind*. Rowman & Littlefield.

Bell, R. L. (2009). *Teaching the nature of science: Three critical questions* [Monograph]. National Geographic School Publishing. http://www.ngspscience.com/profdev/Monographs/SCL22-0449A_SCI_AM_Bell_lores.pdf

Berger, P. L., & Luckmann, T. (1967). *The social construction of reality: A treatise in the sociology of knowledge*. Doubleday.

Canadian Cancer Society. (2024). *Cancer statistics at a glance: Incidence and mortality*. https://cancer.ca/en/research/cancer-statistics/cancer-statistics-at-a-glance

Conrad, P. (1999). Uses of expertise: Sources, quotes, and voice in the reporting of genetics in the news. *Public Understanding of Science, 8*(4), 285–302. https://doi.org/10.1088/0963-6625/8/4/302

Cousineau, M. J. (2015). Mass media as a discursive resource and the construction of engineering selves. *Bulletin of Science, Technology & Society, 35*(1–2), 35–43. https://doi.org/10.1177/0270467615604973

Demirdöğen, B., & Aydın-Günbatar, S. (2021). Teaching nature of science through the use of media reports on COVID-19. *Science Activities, 58*(3), 98–115. https://doi.org/10.1080/00368121.2021.1957757

Dempster, G., Sutherland, G., & Keogh, L. (2022). Scientific research in news media: A case study of misrepresentation, sensationalism and harmful recommendations. *Journal of Science Communication, 21*(01), A06. https://doi.org/10.22323/2.21010206

Einsiedel, E. F. (1992). Framing science and technology in the Canadian press. *Public Understanding of Science, 1*(1), 89–101. https://doi.org/10.1088/0963-6625/1/1/011

Elliott, P. (2006). Reviewing newspaper articles as a technique for enhancing the scientific literacy of student-teachers. *International Journal of Science Education, 28*(11), 1245–1265. https://doi.org/10.1080/10670560500438420

Fairclough, N. (1995). *Media discourse*. Edward Arnold.

Gee, J. P. (2001). Reading as situated language: A sociocognitive perspective. *Journal of Adolescent & Adult Literacy, 44*(8), 714–725. http://www.jstor.org/stable/40018744

Guttmann, A. (2024, April 30). *Daily time spent with selected media among adults in Canada as of 3rd quarter 2023*. Statista. https://www.statista.com/statistics/237478/daily-time-spent-with-media-among-adults-in-canada/

Halkia, K., & Mantzouridis, D. (2005). Students' views and attitudes towards the communication code used in press articles about science. *International Journal of Science Education, 27*(12), 1395–1411. https://doi.org/10.1080/09500690500102912

Jarman, R., & McClune, B. (2002). A survey of the use of newspapers in science instruction by secondary teachers in Northern Ireland. *International Journal of Science Education, 24*(10), 997–1020. https://doi.org/10.1080/09500690210095311

Jarman, R., & McClune, B. (2007). *Developing scientific literacy: Using news media in the classroom*. Open University Press.

Kachan, M. R., Guilbert, S. M., & Bisanz, G. L. (2006). Do teachers ask students to read news in secondary science? Evidence from the Canadian context. *Science Education, 90*(3), 496–521. https://doi.org/10.1002/sce.20113

Leite, M. R. V., Gatti, S. R. T., & Reis, G. (2024). Art and Science: A relationship that permeates the construction of human knowledge about the world. *Science & Education. Advance Online Publication*. https://doi.org/10.1007/s11191-024-00580-6

Lester J. N., & Paulsen, J. (2018). Language-based methodologies for policy and leadership research. In C. R. Lochmiller (Ed.), *Complementary research methods for educational leadership and policy studies* (pp. 57–78). Springer. https://doi.org/10.1007/978-3-319-93539-3_4

Lind, L. (1998). Media education: Learning to reshape the world. *Educational Media International, 35*(1), 4–8. https://doi.org/10.1080/0952398980350103

Luke, A. (1996). Text discourse in education: An introduction to critical discourse analysis. *Review of Research in Education, 21*(1), 3–48. https://doi.org/10.3102/0091732X021001003

MacKenzie, A. H., Lindeman, P., Towler, L., Thomas, W., Muhammad, R., Lorrain, C., Parker, J., Evans, D. L., & Sheldrake, J. (2020). Editor's corner: Scientific media literacy [Special issue]. *The Science Teacher, 87*(5), 6. https://doi.org/10.2505/4/tst20_087_05_6

Macrae, A. (2020). *Discourse analysis*. Oxford University Press. https://doi.org/10.1093/acrefore/9780190201098.013.973

McComas, W. F. (1996). Ten myths of science: Reexamining what we think we know about the nature of science. *School Science & Mathematics, 96*(1), 10–16. https://doi.org/10.1111/j.1949-8594.1996.tb10205.x

Meyer, J. C. (2000). Humor as a double-edged sword: Four functions of humor in communication. *Communication Theory, 10*(3), 310–331. https://doi.org/10.1111/j.1468-2885.2000.tb00194.x

Monteiro, R., & Reis, G. (2018). Animals "Я" us: Egomorphism in/for science and environmental education. *Society & Animals, 28*(5–6), 592–612. https://doi.org/10.1163/15685306-12341526

News Media Canada. (2022). *Canada's newspaper industry in 2022*. https://nmc-mic.ca/about-newspapers/circulation/.

Ontario Ministry of Education. (2022a). *Ontario curriculum, grade 9: Science*. https://www.dcp.edu.gov.on.ca/en/curriculum/secondary-science/courses/snc1w

Ontario Ministry of Education. (2022b). *Ontario curriculum, grades 1–8: Science and technology*. https://www.dcp.edu.gov.on.ca/en/curriculum/science-technology/grades-list

Otte, L., & Beeken, M. (2022). "COVID-19 between myth and science"—Innovative experiments for analyzing fake news inside and outside the chemistry classroom. *Journal of Chemical Education, 99*(5), 1890–1899. https://doi.org/10.1021/acs.jchemed.1c01262

Potter, J. (2003). Discourse analysis and discursive psychology. In P. M. Camic, J. E. Rhodes & L. Yardley (Eds.), *Qualitative research in psychology: Expanding perspectives in methodology and design* (pp. 73–94). American Psychological Association. https://doi.org/10.1037/10595-005

Ratcliffe, M. (1999). Evaluation of abilities in interpreting media reports of scientific research. *International Journal of Science Education, 21*(10), 1085–1099. https://doi.org/10.1080/095006999290200

Reis, G., Jeelani, M., & Brown, A. (2023). "Where is God during a natural disaster?" Potential implications of public discourses of religion for science curricula. In X. Fazio (Ed.), *Science curriculum for the Anthropocene: Curriculum models for our collective future* (Vol. 2, pp. 151–172). Springer. https://doi.org/10.1007/978-3-031-37391-6_8

Reis, G., & Ng-A-Fook, N. (2010). TEK talk: So what? Language and the decolonization of narrative gatekeepers of science education curriculum. *Cultural Studies of Science Education, 5*(4), 1009–1026. https://doi.org/10.1007/s11422-010-9299-x

Retzbach, A., & Maier, M. (2015). Communicating scientific uncertainty: Media effects on public engagement with science. *Communication Research, 42*(3), 429–456. https://doi.org/10.1177/0093650214534967

Roth, W.-M. (2010). An anthropology of reading science texts in online media. *Semiotica, 2010*(182), 409–442. https://doi.org/10.1515/semi.2010.066

Roth, W.-M., & Hsu, P. (2010). Talk about careers in science. *Sense*. https://doi.org/10.1163/9789460913266

Roth, W.-M., Pozzer-Ardenghi, L., & Han, J. Y. (2005). Critical graphicacy: Understanding visual representation practices in school science. *Springer*. https://doi.org/10.1007/1-4020-3376-1

Schäfer, M. S. (2017). How changing media structures are affecting science news coverage. In K. H. Jamieson, D. M. Kahan & D. A. Scheufele (Eds.), *Oxford handbook of the science of science communication* (pp. 50–59). Oxford University Press. https://doi.org/10.1093/oxfordhb/9780190497620.013.5

Shea, N. A. (2015). Examining the nexus of science communication and science education: A content analysis of genetics news articles. *Journal of Research in Science Teaching, 52*(3), 397–409. https://doi.org/10.1002/tea.21193

Statistics Canada. (2023, March 28). *Media consumption in Canada: Are Canadians in the know?* https://www150.statcan.gc.ca/n1/pub/11-627-m/11-627-m2022055-eng.htm

van Dijk, T. (1991). Racism and the press. *Routledge*. https://doi.org/10.4324/9781315682662

Wallace, C. S. (2004). Framing new research in science literacy and language use: Authenticity, multiple discourses, and the "Third Space." *Science Education, 88*(6), 901–914. https://doi.org/10.1002/sce.20024

Wellington, J. (1993). Using newspapers in science education. *School Science Review, 74*(268), 47–52.

Chapter 8
Reflections on the Role of Procedural Knowledge in the Teaching of Science

Elizabeth Vergis

8.1 Introduction

The knowledge explosion of the twenty-first century is unprecedented and has far-reaching effects on the knowledge economy. Traditionally, knowledge was defined by Plato as "justified true belief" (Fine, 2003, p. 5), which means there are three criteria to be met if a statement is to be considered knowledge: it must be justified, true, and believed. Numerous attempts have been made in the past to describe knowledge systematically. One such attempt was made by de Jong and Ferguson-Hessler (1996), whose research was in the field of physics where problem solving, and experimental work have crucial roles. From their work on problem solving, these authors differentiated four types of knowledge: situational, substantive, procedural, and strategic.

Situational knowledge is the knowledge about situations as they typically occur in a particular field or discipline. *Substantive* knowledge concerns itself with facts, concepts, or principles that are applicable within a particular domain. *Procedural* knowledge is the understanding needed to put science into practice; it represents a set of ideas complementary to the substantive knowledge but is more related to the "knowing how" of science (Gott & Duggan, 1995, p. 26). *Strategic* knowledge stipulates the general plan of action in which the sequence of activities to reach a solution is clearly laid down.

Knowledge in science is comprised of facts, information, description, and skills acquired through experience or education (de Jong & Ferguson-Hessler, 1996). Learning science entails mastering the substantive knowledge (content) as well as the procedural knowledge (knowing how). This chapter will examine only substantive

E. Vergis (✉)
University of Alberta, Edmonton, AB, Canada
e-mail: evergis@telus.net

C. D. Tippett and T. M. Milford (eds.), *Seeing Science Through the Eyes of Canadian Teachers and Learners*, Contemporary Trends and Issues in Science Education 54,
https://doi.org/10.1007/978-3-031-88785-7_8

and procedural knowledge because these are the two types of knowledge most often addressed in the teaching and learning of science at the secondary and postsecondary levels.

8.2 Science Knowledge

In this chapter, I will be using data obtained from my doctoral study to substantiate my reflections on the role of procedural knowledge in the teaching of science. I first outline the types of knowledge, nature of evidence, concepts of evidence, and adapted primary literature. Then I address the research questions, methods, and data analyses. Last, I discuss the emergent themes and reflect on the role of procedural knowledge in the teaching of science.

8.2.1 Substantive Knowledge

Substantive knowledge is knowledge about the facts, concepts, theories, and principles that are applicable within a particular domain. Substantive knowledge within a problem-solving domain acts as additional information that solvers can add to the problem and readily use to find the solution. In science, substantive knowledge can be defined as the ideas based on facts, laws, and principles. According to Gott and Duggan (1995):

> Being scientifically literate … requires … that pupils have a sound knowledge base in the major substantive ideas of science and of ideas related to the collection, validation, representation and interpretation of evidence. Without such a repertoire, their ability to think scientifically about a whole range of issues will be impaired. (p. 93).

Therefore, in the survey and interviews conducted for my doctoral work I asked questions based on the substantive knowledge covered in the adapted primary literature as well as the procedural knowledge embedded within the substantive content. A certain level of comfort with the substantive knowledge was necessary to understand the role and significance of the embedded procedural knowledge. In later work, Roberts and Gott (2000) defined "substantive ideas … [as] the concepts which form the underpinning structure of the subjects" (p. 83), for example, ideas such as energy, force, photosynthesis, or solubility.

8.2.2 Procedural Knowledge

Procedural knowledge is the understanding needed to put science into practice. "It is *the thinking behind the doing,*" which includes the understanding of a set of ideas that are complementary to substantive understanding but are "related to the 'knowing how' of science" (Gott & Duggan, 1995, p. 26). "It encompasses the actions or manipulations that are valid within a domain" (de Jong & Ferguson-Hessler, 1996, p. 106). In empirical work, basic experimental skills are needed to succeed, but they are not enough on their own. "There is a higher cognitive dimension, which has been called procedural understanding, necessary for the correct application of those skills" (Roberts & Gott, 1999, p. 22). For example, in a plant growth study, procedural knowledge does not refer to the measuring itself but to the decisions to be made about what to measure, how often to measure, and over what period of time.

Procedural knowledge is often not taught systematically or with rigour. To address this issue, Gott et al. (2002) made a list of approximately 80 concepts of evidence that they classified into 21 categories. The list includes ideas about the uncertainty of data (Buffler et al., 2001) and ideas needed "for understanding measurement and data processing, presentation and analysis which may be considered…essential for understanding evidence" (Roberts et al., 2010, p. 379). It is important to have procedural knowledge incorporated into the K–12 science curriculum so that it is addressed systematically and progressively during the teaching of school science.

8.2.3 Nature of Evidence

Evidence is difficult to define. Philosophers have differing views about evidence. In a dictionary of philosophy, Mautner (1999) defined evidence as "that which provides a ground for a belief or a theory" (p. 184). Popper (1959) asserted that it is evidence that upholds the truth in a proposition, which is particularly true when the proposition in question is based on an experimental observation or experience. Evidence is a concept central to the empirical sciences. Whether or not to believe a scientific hypothesis or theory depends upon the quantity and character of the evidence in its favour.

8.2.4 Concepts of Evidence

Concepts of evidence (CoEs) supply "the underpinning ideas about how evidence can be *collected*, *verified*, *analyzed*, and *interpreted*" [emphasis added] (Roberts & Gott, 2004, p. 11). This term refers to concepts associated with procedural understanding. Skills may include knowing how to use a thermometer, how to draw a graph, how to set up a distillation apparatus, or how to focus a microscope. It is not always recognized that skills have a distinct knowledge base connected with the understanding of scientific evidence. Skills must be exercised within an understanding of ideas such as variables and their manipulation, accuracy, fair testing, and the validity and reliability of data. CoEs are concepts underlying the doing of science in relation to the evidence as a whole. They form a distinct knowledge base akin to the substantive knowledge that has traditionally been seen as the heart of science.

In my doctoral study, I used peer-reviewed scientific literature as a tool to test first-year undergraduate students' understanding of these CoEs. Consequently, it was vital to examine the various ways in which science has been portrayed in the literature and to select the most suitable form for this research project. The purpose of this study, therefore, was to determine whether the pre-instructional understanding that students have of CoEs, as measured by using APL, can be affected by a pedagogical teaching intervention.

8.2.5 Adapted Primary Literature

Adapted primary literature (APL) is a unique genre that was created to enhance the effectiveness of primary scientific literature in the learning of science, especially at the high school level (Falk et al., 2008; Yarden, 2009). It was developed to bridge the distinct gap that exists between the "highly professional nature of research articles and the cognitive level of high school students" (Yarden et al., 2001, p. 191). The genre presents "real science … and to promote important aspects of high school students' scientific literacy that are harder to achieve using textbooks or JRV [Journalistic Reported Version]" (Yarden, 2009, p. 309). In APL, the basic canonical structure of the original primary literature (PL) article is preserved along with the results and supplemental graphics. Alterations make the text simpler but do not modify the text substantially.

The authors of APL are typically science educators and scientists, and the target audience is usually high school science students (Yarden et al., 2001). APL uses mainly the argumentative genre as opposed to the narrative or expository genres usually found in science textbooks (Norris et al., 2009b). Since APL maintains the canonical structure of primary scientific literature (Baram-Tsabari & Yarden, 2005; Yarden et al., 2001), there is enough evidence in its content to support conclusions that are laid down specifically (Falk & Yarden, 2009; Yarden et al., 2001). In APL, science is presented as being uncertain and subject to change, and alterations are expected when new advances are made (Falk & Yarden, 2009; Yarden et al., 2001). It has long been established that "many science curricula … do not take into account the practical reasoning required in scientific knowledge production" (Norris, 1992, p. 196). It is, therefore, plausible to assume that using the argumentative genre prevalent in APL may ensure that scientific reasoning is incorporated into the school curriculum (Yarden, 2009). APL has many promising characteristics:

> It has the ability to introduce complex, authentic, and cutting-edge examples of actual scientific arguments to high school learners. It expands the written text genres available to learners, and assists them in understanding the many different ways that scientific ideas can be conveyed. (Ford, 2009, p. 389).

According to Ford (2009), APL can enhance not only the learning of science content but also the epistemology of science because of the way it is written.

However, not all science educators are convinced that APL is the only way to introduce scientific reasoning into the classroom (Osborne, 2009). Some are of the opinion that APL is secondary literature; therefore, its relevance and accuracy depend on the understanding and writing abilities of the person constructing the APL. Its structure is tied to how the writer understands and interprets what is presented in a journal article as PL. In this regard, an APL can have the same weaknesses as a newspaper article (Norris, Falk, et al., 2008; Osborne, 2009).

Some science educators are of the opinion that instructors, especially at the first-year undergraduate level, should adhere to using PL in their attempts to engage their students as active participants in the learning process (Hoskins et al., 2007; Levine, 2001; Smith, 2001). Instructors could choose articles that are relevant, interesting, and written in a language and style suitable for first-year undergraduate students, although this student group may find the task of reading the PL challenging (Smith, 2001). Moreover, PL is readily available—that is the way scientists communicate with their peers in describing their research and findings (Yarden et al., 2015). Furthermore, science instructors might not have the time or expertise necessary to write good APLs. It could be argued that this task could be assigned to science educators doing research in science education. A good APL depends on choosing good PL, which many science instructors feel students should be exposed to in the first place in order to increase their critical thinking skills.

From a pedagogical perspective, APL seems like a sound approach for teaching science because it uses scaffolding (Vygotsky, 1978) that ensures the learning is anchored on the learner's experience. Such anchoring is a fundamental aspect of the constructivist theory of learning (Driver et al., 1994). Scaffolding, therefore, allows children to accomplish what they could not have achieved on their own because it would have taken too much time and effort. The video lecture used as the teaching intervention in the present study provides the scaffolding needed for the participants to grasp the intricacies of using molecular beacons as the probes for detecting photo-damage in DNA and RNA. It also demonstrates how computer technology is an excellent means of anchoring learning to meaningful experiences.

This research project explored the question: What are some of the factors that influence the teaching and learning of CoEs*?* As the study progressed, the following sub-questions emerged:

1. What difference did the video make to student understanding of substantive and procedural knowledge (i.e., CoEs) addressed in the APL?
2. What did students and instructors think about the APL genre?
3. How well were CoEs conveyed in the APL?
4. What do instructors think about how and when to teach CoEs in chemistry?

8.3 Method

In my study, I explored the understanding of CoEs—students' mental grasp of the ideas about evidence conveyed by these concepts. To examine in depth the understanding of CoEs held by first-year undergraduate students, the research design had to trace the dynamic interrelationship between the APL and the teaching intervention. The theoretical framework of constructivism (Brill et al., 2004) and a qualitative approach to address the research questions met these criteria. However, the data collected were from interviews and an evidence survey administered to only the student participants.

All experiments in educational research share some common features such as offering treatment, monitoring outcomes of the treatment using assigned units, and comparing the outcomes. In the present study, there was at least one treatment (i.e., the teaching intervention), two outcomes that were being measured (i.e., performance on the pretest and the posttest), units of assignment (i.e., marks obtained on the pretest and the posttest), and some comparison (i.e., posttest mark minus the pretest mark) from which a change (i.e., Δ_{diff}) was calculated. Using this method, the Δ_{diff} values were obtained for all participants. Precautions ensured that all other independent variables and conditions were controlled so that the change in marks (Δ_{diff}) could be attributed to the pedagogical treatment.

8.3.1 Construction of the APL from Primary Literature

To prepare the APL, I first identified a suitable piece of primary literature: "Molecular Beacon Probes of Photodamage in Thymine and Uracil Oligonucleotides" by Yarasi et al. (2005). Next, this journal article underwent "a certain degree of processing" (Yarden et al., 2001, p. 191) aimed at simplifying it so that it would be suitable for an audience of first-year undergraduate students. I began writing the APL by dividing the page into two columns: the left column contained the adapted form of the PL maintaining its canonical structure with difficult words underlined, while the right column contained the meaning of the underlined text.

Then, I converted the article's Introduction section into APL and sent this text to two individuals: the chemistry professor who was the first author of the paper (an expert in the field of photochemistry and photobiology), and a professor whose expertise was in science education. These two professors evaluated the APL version of the Introduction and returned it with their corrections. This rigorous process was repetitive and thorough as each section was converted to APL. This process provided a verification that my understanding and interpretation of the PL were checked continuously and monitored constantly by the two experts; it also ensured that any personal bias in writing the APL was kept in check by their expert knowledge. Because the APL was checked and corrected by a team that included the author of the PL, biases were minimized and accuracy was maximized. Finally, I identified the CoEs included in the APL explicitly and implicitly, using Gott's categories of CoEs (Gott et al., 2002) as a conceptual framework.

8.3.2 *Pilot Study*

Science 100 (Integrated Science) is an interdisciplinary course offered in the Faculty of Science at the University of Alberta. The pilot study was conducted in the Science 100 class under the supervision of the instructor; students were given the option of participating in the pilot study or not. All 35 registered students participated; no one opted out in spite of the fact that there were no extra incentives for participation other than the experiential learning offered. Table 8.1 provides the sequence of events for the pilot study.

Over a weekend, students were asked to read the APL and a piece of related secondary literature; both documents were uploaded onto the class's Moodle site. On the following Monday morning, students completed the pretest that was distributed to them as a hard copy during class time. They were then given a lecture on how the chemistry professor and his research group undertook the piece of work described in the APL. Next, students were introduced to CoEs and were specifically taught the CoEs included in the APL paper. Then students completed the posttest in the same class.

I marked the pretests and the posttests immediately. The chemistry professor and I reviewed the questions that were incorrectly answered, checking for ambiguity. Questions that appeared ambiguous or unclear were altered or deleted. We decided to include only the APL in future surveys. The altered pretest and posttest were then put online using SurveyMonkey®. I identified the CoEs in the APL, and the chemistry professor addressed four of the them as he delivered the teaching intervention. This pedagogic intervention was videotaped, resulting in a 32-min video that was uploaded to the class's Moodle site. The video lecture was designed as a first-hand account of the major experiments conducted in the APL, as well as an introduction to CoEs and a few specific ones.

Table 8.1 Order of events in pilot study

	Order of Events	Time Required
1	Read APL and Related Secondary Literature over weekend	2.5 h
2	Sign Consent Form—Monday	10 min
3	Pretest as hard copies—Monday	15–30 min
4	Teaching Intervention as Lecture—Monday	32 min
5	Posttest immediately after the Teaching Intervention—Monday	15–30 min

8.3.3 Study Participants

Participants for the study were enrolled in a first-year undergraduate course, Chemistry 101 or its equivalent, at eight higher education institutions in Alberta. The fact that the data collection sites were a mixture of urban and rural institutions that varied in size from small to large enhanced the diversity of the participants. Initially, I gave a short presentation introducing the research project in each of the 10 Chemistry 101 classes – approximately 300 students in total. All students completed the initial steps in the research sequence as part of their coursework. However, only 61 students consented to the use of their pre- and posttests and to participate in the follow up interviews. Nine of the 10 instructors participated in the follow up interviews.

8.3.4 Research Sequence

The research sequence for this study lasted about 3 weeks. First, all the students in the 10 first-year undergraduate chemistry classes read the APL. Then they were sent a pretest and a demographic survey. When both documents were completed, students were sent the teaching intervention (i.e., the video lecture created from the pilot study). After watching the 32-min video lecture, students were sent a posttest. Subsequently, I invited students to participate in a follow up interview at which point the 61 student participants signed consent forms. Because of time restrictions imposed by the instructors and limitations to my availability, there was sometimes a 2-week gap between the posttest and the follow up interview.

8.3.4.1 Pre-And Posttests

The pretest consisted of 11 questions and the posttest consisted of 13 questions. The tests were designed so that approximately half of the questions examined the substantive knowledge and half examined the procedural knowledge (i.e., the CoEs) addressed in the APL. An example of a question based on the substantive knowledge of the APL is: *From the heating curve of the Molecular Beacons alone, in Fig. 2, how does the fluorescence intensity change as the temperature is increased?* An example of a question based on the CoEs is: *In the experiment shown as Fig. 2, what is the independent (manipulated) variable, and what is the dependent (responding) variable?* The two tests were formatted as online instruments using SurveyMonkey. Since the students took the pretests before the teaching intervention their performance was an appropriate measure of their ability to read and understand a science article. The posttest was intended to reveal how the pedagogical intervention may have contributed to student understanding.

8.3.4.2 APL Intervention

The teaching intervention employed was a video lecture that was sent electronically to each participant. The video depicted a professor in the chemistry department at the University of Alberta who was the principal investigator of the research project described in the APL. In the video, this professor described a few of the experiments and CoEs used explicitly or implicitly in the APL.

8.3.4.3 Interviews

The conceptions that students harboured about the substantive knowledge and the CoE addressed in the APL were probed during semi-structured interviews. These semi-structured interviews were based on 11 pre-established questions for the students and 14 pre-established questions for the instructors. For example, I asked students: *What do you call experiments that we do on the side, to shed light on/strengthen the main experiment?* and I asked instructors: *Can the teaching of Chemistry be made more effective by introducing CoEs?* If the participants' answers warranted further clarification or explanation, I asked additional probing questions.

8.4 Analysis

The pre- and posttests yielded quantitative data, which were analyzed statistically. The interviews yielded qualitative data, which were analyzed using thematic content analysis (Neuendorf, 2019). The first step in analyzing the pre- and posttest responses was to create marking schemes. Despite the differing test lengths, both marking schemes had a total score of 19 points because of multipoint questions. The second step was to calculate the difference between the pre- and posttests, which represented the change in student understanding of the procedural and substantive knowledge addressed in the APL. The third step was to check for significance by conducting a dependent t-test between pre- and posttest scores.

The semi-structured interviews were audio recorded, transcribed, and then coded using codebooks and coding forms. I constructed one codebook for the student interview responses and another for the instructor interview responses by listing the interview questions along with the various possible responses. Each interview response was then coded onto a coding form, and any notes about the responses were recorded at the bottom of the form. I then analyzed the coding forms to identify themes, the central notions that emerged from the responses.

8.5 Results and Discussion

Pre-and posttest scores provided the quantitative results from my study. Mean scores for all participants before and after the APL intervention were compared for differences and analyzed for significance. The qualitative results came from the analysis of the interview responses, where four main themes emerged: the video as a teaching intervention, grappling with the APL, student understanding of CoEs, and how and when to teach CoEs. In the remainder of this section, which is structured around the research questions, students' and instructors' comments are identified with an S or I and a numeral to maintain confidentiality.

8.5.1 What Impact Did the Video Have on Understanding CoEs?

Research Question 1 was *What difference did the video make to student understanding of substantive and procedural knowledge (i.e., CoEs) addressed in the APL?* Both quantitative and qualitative results suggested that the video impacted student understanding of CoEs in a number of ways.

The mean score for the pretest was 8.74 (SD = 2.49) while the mean score for the posttest was 10.39 (SD = 2.64). The change in understanding (i.e., Δ_{diff}) was found to be positive for 85% of participants, indicating that these students scored better on the posttest than on the pretest. The average change in scores was 1.92 points (SD = 2.23). The largest positive change was 6.25 and the largest negative change was -3.25. and found significant differences ($t = 6.74$, $df = 60$, $p < 0.001$). The positive change (i.e., Δ_{diff} with positive values) was obtained when a student's score on the posttest was higher than the score on the pretest. This implied that the teaching intervention improved the students' understanding of the APL and of the Concepts of Evidence embedded in it. The negative change (i.e., Δ_{diff} with negative values), on the other hand, indicated that a student's score on the posttest was lower than the score on the pretest. This rarer phenomenon suggested that the teaching intervention did not promote a deeper understanding of the content of the APL and of the Concepts of Evidence embedded within it.

With regards to qualitative results, the first theme that emerged from the thematic analysis centred on the video as a teaching intervention. More than four-fifths of the students found the video helpful, with more than half of them readily referring to it as excellent or very useful in helping them to comprehend the APL and develop an understanding of the CoEs entrenched in the paper. According to S5, "[The video] made a big difference. That's when I understood." S16 found the video very useful because it was much more in depth than the APL. Students found the video useful because it was easy to follow, it covered the background information very clearly, it summarized the main points, it was well organized, and the information contained in it was useful. Some misunderstandings arising from the APL were explained in the video. The pace of teaching in the video was not rushed and students found it was ideal to just sit and watch.

Roughly two-fifths of the students appreciated the drawings and flowcharts presented on the board during the video because they were both useful and informative. The auditory and visual impact of the video in teaching about the content knowledge and the CoEs were found to be effective by S30, who claimed:

> The video was the most positive aspect in that it relayed and provided auditory teaching. It provided visual imagery. And so, rather than just sitting and reading, it provided different stimuli, which I feel increases the retention and ability to understand. I guess it's like someone breaking it down and explaining it. … Certainly, maybe it is repetitious, because they are presenting the same data, but it allows for the mind to process it, in a different way, in different pathways. Ears versus the eyes, it is very, very useful.

Almost three-quarters of the students said that the most beneficial part of this research activity was the video. S30 summed it up, saying "The video was awesome." For many students, the difference made by the video was so remarkable that "it was like the light came on" (S26). "I thought it was excellent," remarked S40. This sentiment was echoed by S9, who claimed that out of the whole research activity, "I liked the video best."

Most instructors believed that the teaching intervention would promote student understanding of the CoEs. The visuals and anecdotal explanations were identified as successfully linking empirical evidence to underlying scientific theories. One instructor vouched that the teaching intervention would certainly further the students' learning of the substantive and procedural knowledge (i.e., CoEs) contained in the APL. When asked whether the teaching intervention in the form of the video can make a difference, I5 responded:

> I think so. I like being in front of .... I like teaching live so, I think that's one way, but I don't think the only way, ... so having something different and especially this day and age of students, they are more video, audio .... so the idea of them, you know, watching something and paying attention to it is probably, they might find that more exciting than a live teaching. So, I would imagine it certainly was effective. Yeah.

I7 also commented favourably about the teaching intervention, saying "I think it will make a huge difference" adding that although there was only one video, it would likely further the students' understanding of both substantive and procedural knowledge embedded in the APL.

8.5.2 What Did Participants Think About the APL Genre?

Research Question 2 was *What did students and instructors think about the APL genre?* This question was answered by the second theme that emerged from the thematic analysis, which dealt with the students grappling with the APL as they tried to make sense of it. Many students had to read the APL once, then come back and read it a second or even a third time, as they struggled to understand what was happening in some of the experiments described.

Four students found it discouraging that though they read the APL, they failed to fully understand it. For example, S1 said that it took a while to realize that they had to note down points as they read the APL and refer back to them as they answered the questions. However, more than half of the students reported to have enjoyed reading the APL, although sometimes this enjoyment came with reservations. For example, S14 enjoyed reading the APL but noted "some of it was hard to understand and read." S29 commented:

> It took some close critical reading, crash [focussed]reading once or twice. It wasn't too bad, because I hadn't seen papers like that so far, in my research or in my classes … so this was a kind of nice and [the best part was having] the definitions on the side. So, it wasn't too bad.

Some of the instructors were concerned about the content of the APL, pointing to what they considered to be omissions. For example, they noted arbitrary units used in the figures, a lack of clarity concerning the number of trials in the experiments, and a missing justification for selecting the two oligonucleotides used in the experiments. However, upon further investigation, I learned that the authors of the PL had conformed to the graphing conventions for their work, and that the PL was equally unclear regarding the number of trials and the selection of molecules resembling DNA and RNA.

8.5.3 How Well Were CoEs Conveyed in the APL?

Research Question 3 was *How well were CoEs conveyed in the APL?* and this question was answered by the third theme that emerged from the thematic analysis, which centered on student familiarity with CoEs and its impact on content knowledge. Some of the CoE-related challenges identified during interviews with students included dealing with anomalous data, designing a fair test, manipulating independent variables, and identifying scientific purpose of a series of experiments. The students struggled with these concepts as shown in the following examples.

When discussing data that deviates from the general trend, S12 explained that if there were multiple occurrences of anomalous data, the solution would probably be to redo the experiment, but added, "Aside from that I could just call it error, if I had just one outlier." When talking about designing a fair test, S14 said, "You can't just prove it actually true or false. You have to [do] all those controls." This student failed to recognize that only one independent variable is altered at one time. When asked about independent variables, the majority of the students knew that only one independent variable could be changed at a time, although when asked if two independent variables could be changed at the same time S37 said, "You probably can do the procedure, but it will be more difficult." Fewer than half of the students could identify that the purpose of the experiments that they had read about was to determine optimal conditions for using molecular beacon probes to measure photodamage. The rest of the students gave responses including "I don't know" (S5), "They are trying to find the energy level which affects the DNA" (S14) and "To see if photo light gives a direct reason why things are the way they are…trying wavelengths" (S11).

From the instructor point of view, "the thinking gets kind of lost in the content" I5. This sentiment suggests that the thinking associated with the CoEs or procedural knowledge gets somewhat submerged in the substantive or content knowledge. It seems clear from both student and instructor interview responses that in order to recognize and understand the role of CoEs in the APL, the reader needed to have a good grasp of the substantive knowledge covered.

8.5.4 What Do Instructors Think About Teaching CoEs?

Research Question 4 was *What do instructors think about how and when to teach CoEs in chemistry?* This question was answered by the fourth theme that emerged from the thematic analysis, which dealt with implicit versus explicit teaching of CoEs and the appropriate level for teaching CoEs. Instructors' interviews suggested that there were two schools of thought about teaching CoEs. One school of thought – in the minority – was that chemists acquire the necessary procedural knowledge implicitly by working on projects and ideas, struggling with them, and dealing with complicated issues. "They don't get it by your standing there and telling them, 'These are the Concepts of Evidence'," said I4. According to this instructor, a better way of teaching these CoEs would be to keep them "below the surface" and allow the students to explore them for themselves. I4 was convinced that was the way to make the students really learn these CoEs and further added:

> I think it's better to teach those concepts in a 'soft' way, as undertones to your education, not as explicit ideas. I think that's straight-jacketed. But maybe I'm just queer. That's not how chemists would work. I am not going to go through a list and say, 'Oh, look at all my Concepts of Evidence.' You know what the evidence for a particular argument is. You know what errors are. You know how errors work. You get that through experience.

This instructor seemed to advocate that students struggling with these CoEs should work on their own, without any formal, explicit teaching of these constructs. The second school of thought was that these CoEs have to be taught explicitly. This idea was echoed by eight of nine instructors who were in total agreement and only wondered when this teaching should occur.

Most instructors were of the opinion that CoEs should be taught as early as possible in the first year of university. I5 reiterated:

> I think it would be better if we emphasize more of that from the word 'go.' I can't imagine getting to fourth year and wanting to go to Grad School and not being able to collect data properly and so I think it needs to be started in first year.

However, I2 had a very interesting comment, insisting that it should be at the first-year undergraduate level and not in high school:

> I have a problem with high school trying to get their paws into all sorts of things that are high level, and they don't get the basics. I think that high school needs to go back to teaching the basics, and let postsecondary people do, you know, the higher-level things. High schools can teach the basics .... I'd rather they learnt their chemistry a little better rather than be able to have this kind of critical thinking at this level. They don't have enough body of knowledge to do any practical work with this kind of work.

According to I2, high school teaching should be limited to the basics. Higher level thinking tasks, such as critical thinking skills and students judging their work and that of others, should be left to the postsecondary institutions and should not even be attempted at the high school level. I2 was certain that the explicit teaching of CoEs should commence only at the first-year undergraduate level and that high school teachers should equip their students with a body of knowledge and instill in them an ability to do mathematics. I2 was quick to point out:

> They know how to work in groups. They know how to do all that kind of stuff, but the body of knowledge, they don't have it in them. They can look it up in a data booklet or on the computer. But they don't actually know anything.

I2 expected students leaving high school to have a solid and fundamentally basic repertoire of knowledge to fall back on when faced with decisions in chemistry classes or laboratories; this repertoire, according to I2, seems to be lacking in students graduating from high schools today.

8.5.5 Conclusions

My initial question was *"Can Concepts of Evidence be taught or learnt and, if so, what are the factors that influence their teaching and learning?"* The qualitative findings appear to answer this question in the affirmative because the majority (85%) of the students who participated in the evidence survey showed an improvement in their performance on the posttest. From their pretest scores, it was evident that these students did not come to this research activity as blank slates but that they had pre-instructional conceptions, which may be either opposed to or in line with the scientific view. When these pre-instructional conceptions do not comply with the scientific view, they are referred to as misconceptions or alternate conceptions. One role of a teaching intervention is to address such misconceptions.

According to Gott and Duggan's model of science (1995), knowledge itself can be differentiated into two main types: substantive and procedural. Constructivism, on the other hand, treats knowledge as one entity. In the model, facts contribute to substantive understanding while skills furnish the procedural understanding. It is by the interaction of both substantive and procedural knowledge that cognitive processes are stimulated to solve problems. The arrow from substantive understanding to the cognitive processes is unidirectional—so, too, is the arrow linking procedural understanding to the cognitive processes. In addition, there is no direct connection between substantive knowledge and procedural knowledge. Recent work (Roberts et al.,

2010) sheds light on the importance of substantive and procedural understanding in performing open-ended science investigations. The authors took the stance that substantive and procedural understanding are "two connected but distinct types of knowledge" (p. 377) and emphasized the need at times to address procedural knowledge specifically and separately from substantive content. My work also suggests that a certain degree of substantive knowledge is required for a proper understanding of procedural knowledge and that understanding procedural knowledge thoroughly promotes substantive knowledge as well.

Therefore, I would modify Gott's model of science by inserting a bidirectional arrow connecting substantive knowledge and procedural knowledge. This modification would indicate that substantive knowledge promotes procedural understanding and vice versa. I have shown this modification clearly in Fig. 8.1. My study seems to suggest that there is a mutually enhancing, iterative relationship between substantive knowledge and procedural understanding.

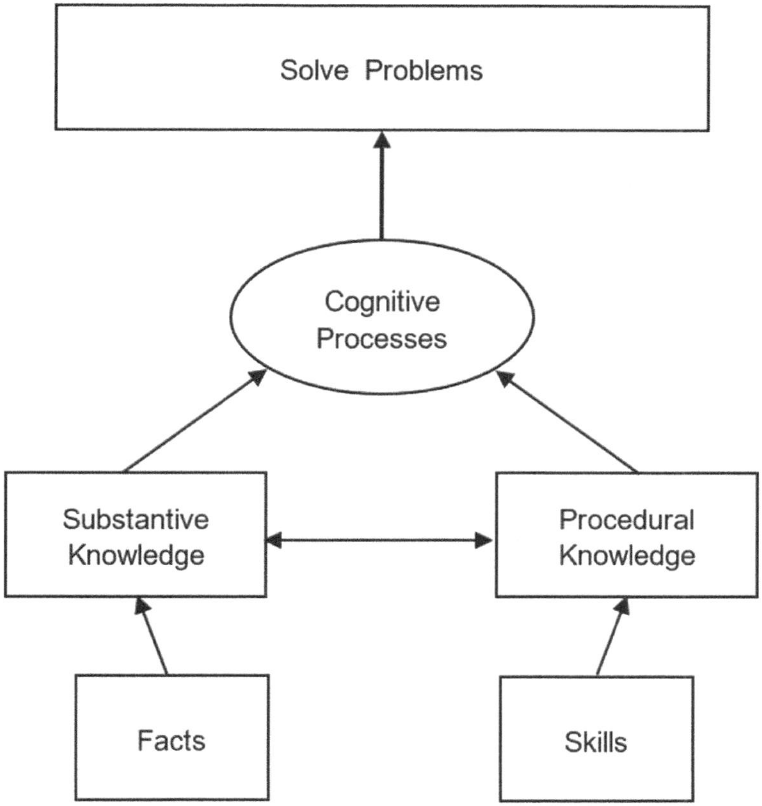

Fig. 8.1 Modified model for knowledge in science (adapted from Vergis, 2017). *Note.* Based on Gott and Mashiter (1991)

One of the greatest criticisms by science educators (G. P. Thomas, personal communication, June 29, 2012) levelled at the classification of CoEs by Gott et al. (2002) is that there are too many categories and subcategories. Unfortunately, this complexity has not been documented or addressed in the science education literature; what I describe is an original contribution. I have adapted a scheme that designates each CoEs category as one of four types. This overall designation is presented in Fig. 8.2. If one wants to consider "Measurement," for example, one can look up the seven CoEs under that group and decide which one to choose. The access to and the recall of the appropriate CoEs are, therefore, made much easier, more systematic, and less cumbersome.

The results from this study suggest that procedural knowledge should be taught *piecemeal*, starting as early as Grade 4 and extending through junior secondary and secondary school. Some of the more complex CoEs could be taught piecemeal starting as early as first-year undergraduate and extending to the senior third and fourth years postsecondary. When implemented, undergraduate students taking science courses will be able to see threads of procedural understanding running through all the science disciplines, which may enhance student engagement. To enable this progression, the curriculum should be developed so that the category and depth of procedural knowledge, or skills, are specified clearly and developed as precisely for every topic just as for substantive knowledge. In addition, the testing and assessment of procedural knowledge should be made mandatory. Assessment could be implemented more effectively if the procedural knowledge is sequenced with the related substantive knowledge in the science curriculum.

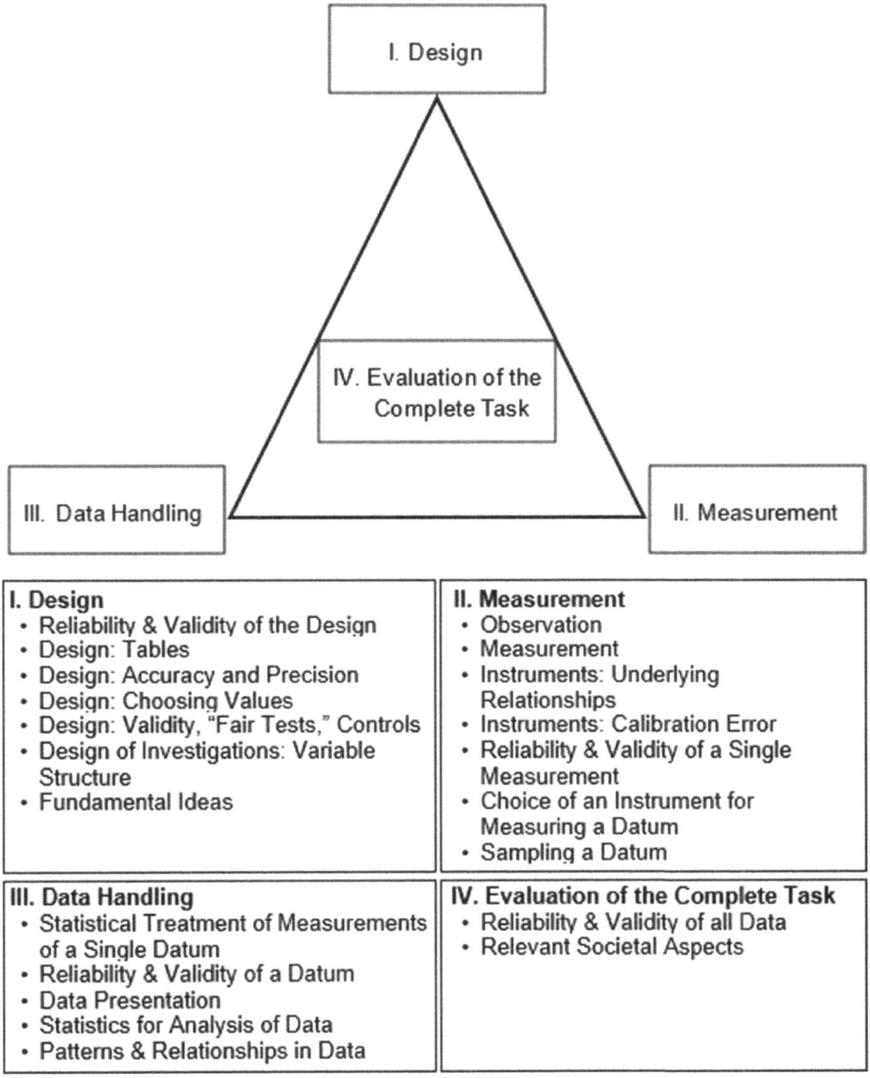

Fig. 8.2 Categories ($N = 21$) of Concepts of Evidence designated to four groups (adapted from Vergis, 2017)

References

Baram-Tsabari, A., & Yarden, A. (2005). Text genre as a factor in the formation of scientific literacy. *Journal of Research in Science Teaching, 42*(4), 403–428. https://doi.org/10.1002/tea.20063

Brill, G., Falk, H., & Yarden, A. (2004). The learning processes of two high-school biology students when reading primary literature. *International Journal of Science Education, 26*(4), 497–512. https://doi.org/10.1080/0950069032000119465

Buffler, A., Allie, S., & Lubben, F. (2001). The development of first year physics students' ideas about measurement in terms of point and set paradigms. *International Journal of Science Education, 23*(11), 1137–56. https://doi.org/10.1080/09500690110039567

Çalik, M., Kolomuç, A., & Karagölge, Z. (2010). The effect of conceptual change pedagogy on students' conceptions of rate of reaction. *Journal of Science Education and Technology, 19*(5), 422–433. https://doi.org/10.1007/s10956-010-9208-9

de Jong, T., & Ferguson-Hessler, M. G. M. (1996). Types and qualities of knowledge. *Educational Psychologist, 31*(2), 105–113. https://doi.org/10.1207/s15326985ep3102_2

Driver, R., Asoko, H., Leach, J., Scott, P., & Mortimer, E. (1994). Constructing scientific knowledge in the classroom. *Educational Researcher, 23*(7), 5–12. https://doi.org/10.3102/0013189x023007005

Falk, H., Brill, G., & Yarden, A. (2008). Teaching a biotechnology curriculum based on adapted primary literature. *International Journal of Science Education, 30*(14), 1841–1866. https://doi.org/10.1080/09500690701579553

Falk, H., & Yarden, A. (2009). "Here the scientists explain what I said." Coordination practices elicited during the enactment of the results and discussion sections of adapted primary literature. *Research in Science Education, 39*(3), 349–383. https://doi.org/10.1007/s11165-008-9114-9

Fine, G. (2003). *Plato on knowledge and forms: Selected essays*. Oxford University Press.

Ford, D. J. (2009). Promises and challenges for the use of adapted primary literature in science curricula: Commentary. *Research in Science Education, 39*(3), 385–390. https://doi.org/10.1007/s11165-008-9115-8

Gott, R., & Duggan, S. (1995). *Investigative work in the science curriculum*. Open University Press.

Gott, R., Duggan, S., & Roberts, R. (2002). *Concepts of evidence: The Thinking Behind the Doing (GCSE version)*. Durham University. http://cofev.webspace.durham.ac.uk/wp-content/uploads/sites/299/2022/04/CofEv_PDF_GCSE.pdf

Gott, R., & Mashiter, J. (1991). Practical work in science -a task-based approach? In B. E. Woolnough (Ed.), *Practical science* (pp. 53–67). Open University Press.

Hoskins, S. G., Stevens, L. M., & Nehm, R. H. (2007). Selective use of the primary literature transforms the classroom into a virtual laboratory. *Genetics, 176*(3), 1381–1389. https://doi.org/10.1534/genetics.107.071183

Hyland, K. (1996). Writing without conviction? Hedging in science research articles. *Applied Linguistics, 17*(4), 433–454. https://doi.org/10.1093/applin/17.4.433

Levine, E. (2001). Reading your way to scientific literacy: Interpreting scientific articles through small-group discussions. *Journal of College Science Teaching, 31*(2), 122–125. http://www.jstor.org/stable/42991325

Mautner, T. (Ed.). (1999). *The Penguin dictionary of philosophy*. Penguin Books.

Neuendorf, K. A. (2017). *The content analysis guidebook* (2nd ed.). SAGE.

Neuendorf, K. A. (2019). Content analysis and thematic analysis. In P. Brough (Ed.), *Advanced research methods for applied psychology: Design, analysis and reporting* (pp. 211–223). Routledge.

Norris, S. P. (1992). Practical reasoning in the production of scientific knowledge. In R. A. Duschl & R. J. Hamilton (Eds.), *Philosophy of science, cognitive psychology, and educational theory and practice* (pp. 195–225). State University of New York Press.

Norris, S. P., Falk, H., Federico-Agraso, M., Jiménez-Aleixandre, M. P., Phillips, L. M., & Yarden, A. (2009). Reading science texts—Epistemology, inquiry, authenticity—A rejoinder to Jonathan Osborne. *Research in Science Education, 39*(3), 405–410. https://doi.org/10.1007/s11165-008-9118-5

Norris, S. P., Macnab, J. S., Wonham, M., & de Vries, G. (2009). West Nile virus: Using adapted primary literature in mathematical biology to teach scientific and mathematical reasoning in high school. *Research in Science Education, 39*(3), 321–329. https://doi.org/10.1007/s11165-008-9112-y

Osborne, J. (2009). The potential of adapted primary literature (APL) for learning: A response. *Research in Science Education, 39*(3), 397–403. https://doi.org/10.1007/s11165-008-9117-6

Popper, K. R. (1959). *The logic of scientific discovery*. Hutchinson & Co. (Original work published 1935)

Roberts, R., & Gott, R. (1999). Procedural understanding: Its place in the biology curriculum. *School Science Review, 81*(294), 19–25.

Roberts, R., & Gott, R. (2000). Procedural understanding in biology: How is it characterized in texts? *School Science Review, 82*(298), 83–91.

Roberts, R., & Gott, R. (2004). A written test for procedural understanding: A way forward for assessment in the UK science curriculum? *Research in Science and Technological Education, 22*(1), 5–15. https://doi.org/10.1080/0263514042000187511

Roberts, R., Gott, R., & Glaesser, J. (2010). Students' approaches to open-ended science investigation: The importance of substantive and procedural understanding. *Research Papers in Education, 25*(4), 377–407. https://doi.org/10.1080/02671520902980680

Smith, G. R. (2001). Guided literature explorations: Introducing students to the primary literature. *Journal of College Science Teaching, 30*(7), 465–469. 9. http://www.jstor.org/stable/42992182

Vartalla, T. (1999). Remarks on the communicative functions of hedging in popular scientific and specialist research articles on medicine. *English for Specific Purposes, 18*(2), 177–200. https://doi.org/10.1016/S0889-4906(98)00007-6

Vergis, E. (2017). *Using Adapted Primary Literature to test the understanding of Concepts of Evidence in chemistry held by first year university undergraduate students.* [Doctoral dissertation, University of Alberta]. Education and Research Archive. https://doi.org/10.7939/R3SQ8QX5G

Vygotsky, L. S. (1978). *Mind in society: The development of higher psychological processes.* Harvard University Press.

Yarasi, S., McConachie, C., & Loppnow, G. R. (2005). Molecular beacon probes of photodamage in thymine and uracil oligonucleotides. *Photochemistry and Photobiology, 81*(2), 467–473. https://doi.org/10.1111/j.1751-1097.2005.tb00209.x

Yarden, A. (2009). Reading scientific texts: Adapting primary literature for promoting scientific literacy. *Research in Science Education, 39*(3), 307–311. https://doi.org/10.1007/s11165-009-9124-2

Yarden, A., Brill, G., & Falk, H. (2001). Primary literature as a basis for a high-school biology curriculum. *Journal of Biological Education, 35*(4), 190–195. https://doi.org/10.1080/00219266.2001.9655776

Yarden, A., Norris, S. P., & Phillips, L. M. (2015). *Adapted primary literature: The use of authentic scientific texts in secondary schools.* Springer.

Chapter 9
Cognitive Presence Indicators in Undergraduate Science Asynchronous Problem-Based Learning Discussions

Michael-Anne Noble

9.1 Introduction

In this chapter I present a novel comparison of the online discussions of undergraduate environmental science students in distance or blended ecotoxicology courses during the same online problem-based learning (PBL) case. This fourth-year course is a capstone course in an interdisciplinary Bachelor of Science in Environmental Science program. Online PBL is of interest due to the prevalence of online education in the face of challenges such as the COVID pandemic and increased demands for education to be accessible remotely. Royal Roads University (RRU), where this study took place, has a Learning, Teaching, and Research Model (LTRM) with a strong emphasis on distance (online) and blended (face-to-face combined with online) educational approaches including the use of PBL (RRU, n.d.). This model emphasizes learning that is outcomes-based, technology-enhanced, experiential, authentic, community-based, team-based, integrative, applied, engaged, action research-focused, supportive, and flexible. I used the Community of Inquiry (CoI) framework (Garrison et al., 2001) to examine cognitive presence in the context of online PBL activities in asynchronous discussion forums.

9.1.1 Community of Inquiry Framework

The CoI framework (Garrison et al., 2001) was developed to examine asynchronous online discussions; it has a straightforward coding scheme with validated indicators that can be used to examine higher-order thinking skills as expressed by cognitive

M.-A. Noble (✉)
Royal Roads University, Colwood, BC, Canada
e-mail: mickie.noble@royalroads.ca

C. D. Tippett and T. M. Milford (eds.), *Seeing Science Through the Eyes of Canadian Teachers and Learners*, Contemporary Trends and Issues in Science Education 54,
https://doi.org/10.1007/978-3-031-88785-7_9

presence indicators. Garrison (2017) described an online education experience as being made up of three intersecting parts: cognitive presence, teaching presence, and social presence (see Fig. 9.1). Cognitive presence is "the exploration, construction, resolution and confirmation of understanding through collaboration and reflection in a community of inquiry." (Garrison, 2007, p. 65). CoI cognitive presence indicators can be operationalized with four progressive and hierarchical phases—triggering event, exploration, integration, and resolution (Garrison, 2017)—with progress through the phases indicating higher levels of cognitive presence. In the triggering event phase, a problem is recognized. In the exploration phase, the nature of the problem is characterized then relevant information is explored. In the integration phase, meaning is constructed and ideas are tested. In the resolution phase, a solution may be implemented and examined or tested through further discussion. A resolution may lead to additional triggering events for complex problems, thus re-starting the exploration phase. An asynchronous online environment can support cognitive presence because individuals have time to engage in reflection and to respond when they are ready (Meyer, 2003).

Teaching presence is "the design, facilitation, and direction of cognitive and social processes for the purpose of realizing personally meaningful and educationally

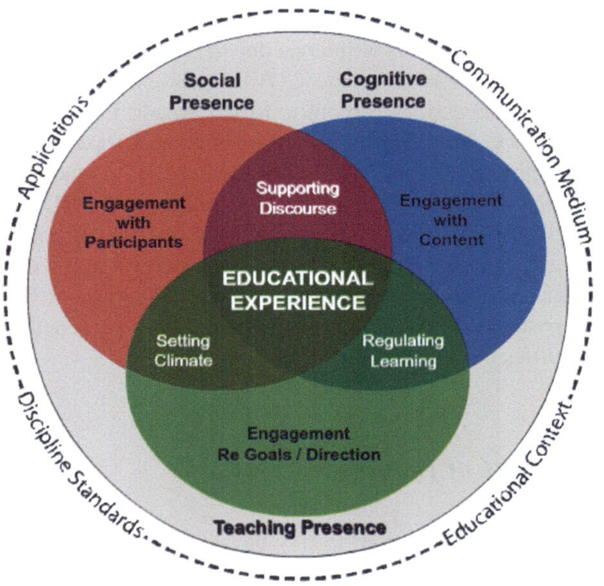

Note. From "Community of Inquiry Framework," by D. R. Garrison, M. Cleveland-Innes, and N. D. Vaughan, n.d. (https://coi.athabascau.ca). Reprinted with permission.

Fig. 9.1 Community of Inquiry framework for an online educational experience. *Note* From "Community of Inquiry Framework," by D. R. Garrison, M. Cleveland-Innes, and N. D. Vaughan, n.d. (https://coi.athabascau.ca). Reprinted with permission

worthwhile learning outcomes" (Garrison, 2017, p. 27). Teaching presence has three indicators: design and organization, facilitating discourse, and direct instruction. The design and organization indicator includes the design and layout of course assignments, readings, and other resources/activities. The facilitating discourse indicator includes encouraging student interaction and discussion. The direct instruction indicator includes providing information/advice and clarifying ideas (Anderson et al., 2001; Vaughan et al, 2013). Teaching presence indicators often include scaffolding—the temporary supports provided to guide students through problems that they would not otherwise be able to complete (Schmidt et al., 2011).

Social presence is "the ability of participants to identify with the group or course of study, communicate purposefully in a trusting environment, and develop personal and affective relationships progressively by way of projecting their individual personalities" (Garrison, 2017, p. 41). The three indicators for social presence are affective communication, open communication, and cohesion. Affective communication includes such things as expressing emotions, using humour, or telling the team something about yourself or your feelings (self-disclosure) (Garrison, 2017, p. 45). Open communication includes referring to or quoting from someone else's message, complimenting someone's post, agreeing with or replying to someone's post. Cohesion includes referring to others by name, using inclusive pronouns, and using salutations to start a message (Garrison, 2017, p. 46). These social indicators can help to build and maintain the learning community of the group.

The CoI framework was developed to examine asynchronous online discussions and has a straightforward coding scheme for cognitive presence indicators, which were my focus. CoI has been confirmed by factor analysis (Arbaugh, 2007; Garrison et al., 2004); it has been used by multiple researchers (e.g., Heckman & Annabi, 2005; Meyer, 2003) and as a framework by RRU instructional designers (RRU, n.d.), justifying it an appropriate framework for this investigation of online PBL.

9.1.2 What is Problem-Based Learning (PBL)?

PBL is a team-based learning method that focuses students on solving real-world problems (Barrows, 1996, 2002; Hmelo-Silver, 2004). These authentic problems provide a focus and stimulus for learning as well as relevant contexts in which to apply that learning (Barrows, 1996; Hmelo-Silver, 2004; Koschmann et al., 1994). As online education becomes more common, the question of how PBL can be adapted from face-to-face to online environments becomes more important and particularly urgent in the context of situations such as the COVID-19 pandemic.

Walker and Leary (2009) suggested that "PBL may do best ... when assessment is at the application rather than the conceptual level" (p. 28). Given the multidisciplinary nature of ecotoxicology and the applied nature of the RRU BSc programs, PBL is well suited for this course. Moreover, given RRU's increasing emphasis on online education and the use of geographically distributed teams for course projects, PBL in an online format is worthy of further exploration. I teach an ecotoxicology course in both distance and blended formats that includes a PBL case. Distance cohorts work entirely online while blended cohorts experience mostly face-to-face instruction; the PBL case for both formats occurs asynchronously on Moodle™ (the learning management system used at RRU).

In order to develop this PBL case, I reviewed a number of real-world problems (e.g., Exxon Valdez, Chernobyl) that would potentially encourage students to investigate some key course concepts. However, none of the problems led students to investigate all key concepts, and solutions were readily available. I decided to develop a case that consisted of a sequence of four cumulative case problems designed around a fictitious community that contained various industries. Students would work through the problem sequence in teams that varied between six and eight members. Each problem was posed in the form of a narrative that identified the role of the team (e.g., a member of a local nongovernmental organization or the town's environmental officer) in order to provide a more authentic context for the case. In these problems I included scaffolds that described some of the necessary concepts or content that the team was required to address in its case report. Teams had to submit a report summarizing their solution to one problem before they received information about the next problem. As in the real world, decisions made during the earlier problems determined the choices that were available in the latter problems and resulted in a variety of possible solutions.

For each of the four PBL case problems, each team had a separate discussion forum that was visible only to the team members and the instructor. One or two students from each team acted as moderators for each case problem, and this role was rotated to new students for each case problem. It was a course requirement that everyone on the team had to have a turn as the moderator during one of the four problems. Moderators saw this as their chance to decide how the discussion would be structured and to practice their leadership skills. In consultation with the team, moderators had the ability to decide who could start new threads (i.e., moderator only or everyone), if a template was to be used, and what the soft deadlines were. They also assigned roles and research areas; ensured that all the required information was collected, vetted, and assembled into the agreed-upon format; and prompted team members for participation, additional information, or justification as needed. Usually, the moderator circulated a draft to the group for comments, suggestions, and editing prior to submitting a summary report that presented the team's solution to the PBL case problem. The PBL portion of the course accounted for one-third of the course work and occurred in asynchronous online discussion forums on Moodle regardless of the instructional format used for the rest of the course.

9.2 Methods

My aim was to compare distance and blended teams with respect to their online activity as demonstrated by CoI indicators of cognitive presence. My initial observations as the course instructor indicated that their activity seemed to be different in terms of level and content, and I was interested in what might be causing the differences. Meyer (2003) described the categories of exploration, integration, and resolution in the CoI framework as exhibiting higher-order thinking skills because such posts indicate that learners are involved in activities that go beyond just posting information. The CoI framework was used in this study to examine cognitive presence in the context of online PBL activity in asynchronous discussion forums. This mixed methods study represents my exploration of the CoI cognitive presence indicators in teams' asynchronous PBL discussions. The research question that guided this study was:

- What patterns emerge when using CoI cognitive indicators to assess the asynchronous discussions of distance and blended cohorts during online PBL?

I examined the asynchronous PBL case discussion forums for four distance cohorts of 108 students in the BSc Environmental Management program and four blended cohorts of 144 students in the BSc Environmental Science program, all of whom were enrolled in Ecotoxicology 407. There were three methods used for this work. Quantitatively, the academic outcomes for distance and blended students were compared using independent samples *t* tests for the scores achieved on the scenario included on the final examination as well as the scores for the PBL case problems. Qualitatively, content analysis and thread maps—graphic representations of how asynchronous online discussions occur over time (Pena-Shaff & Nicholls, 2004)—were used. All posts were subjected to content analysis to decide which CoI indicator the post corresponded to. Finally, thread maps were constructed by overlaying the CoI coding over visual representations of the timing and arrangement of posts within each asynchronous online discussion forum.

9.2.1 Data Collection and Analysis

In this section I outline the data collection and analysis procedures for the academic outcomes, content analysis of the discussion posts, and creation of thread maps. Statistical analysis was done using IBM SPSS 19.0. QSR NVivo10 was used for coding of the posts as well as for quantifying the type of posts.

9.2.1.1 Academic Outcomes

Overall course grades were not used to compare the distance and blended groups since the deliverables for the non-PBL parts of the course differed. However, a direct comparison of academic outcomes was possible using the scores for the scenario-based section of the final examination as well as the combined grades for the four PBL case problems. The scenario question on the final examination resembled the PBL case problems used in the course. My decision to use these marks for comparison is supported by the PBL literature that shows students' examination performance is better when the examination is similar to the PBL problems (scenario based) than on general examinations (Dochy et al., 2003; Gijbels et al., 2005; Strobel & van Barneveld, 2009). The distance and blended mean scores on the PBL-like portion of the final examination were marked out of 110 possible points and were compared using an independent samples *t* test. A second independent samples *t* test was used to compare the combined grades for the four PBL case problems between the distance and blended groups. Scoring was based on a total of 25 points for the four problems.

9.2.1.2 Content Analysis and CoI Coding of Discussion Posts

To collect posts, I downloaded the PBL discussions for each team as compiled transcripts from the Moodle archives along with any files that were attached to discussion posts (e.g., Excel, Word, PDF, Paint, PowerPoint). The transcripts were converted to PDF to maintain the appearance and formatting of the posts; this process was important because individual posts could be readily identified. I chose the post as the unit of analysis for this study because (a) it is a discrete contribution as determined by the message author (Rourke et al., 2001) and (b) it allowed comparison to other studies that used posts as the unit of analysis (e.g., Garrison et al., 2001; Meyer, 2003).

Initially, I categorized each post in each discussion forum as either triggering, exploration, integration, resolution, or other. Posts coded as triggering were ones in which students asked questions that took the discussion in new directions, proposed tangents to the discussion, or presented background information leading into a question. Posts coded as exploration included contradictions, personal narratives, opinions, and brainstorming. Posts coded as integration built on ideas, substantiated points, provided evidence, integrated ideas from several different sources, and discussed or referred to other members' posts. Posts coded as resolution were ones in which students tested and defended possible solutions or discussed applications to real-world situations. Posts coded as other did not fit into the four cognitive presence categories; I had noticed that many of these posts either served to organize discussions or negotiate how the team was to work together (e.g., action plans, organizational templates, volunteering for tasks).

These posts did not fit into the social presence indicators (i.e., open communication, group cohesion, or affective expression) or teaching presence indicators (i.e., design and organization, facilitating discourse, or direct instruction) parts of the CoI model (Garrison & Arbaugh, 2007; see Fig. 9.1). The posts might have been coded *design and organization* had they been instructor posts; however, because they were student posts, I added a new category of *organization* within the cognitive presence to capture these student contributions. Table 9.1 shows pseudoexamples of posts that were coded to the various cognitive presence indicators. Pseudoexamples are provided rather than direct quotes because the ethics approval for this study required that, due to the size of the teams within each cohort and the identification of cohorts, only limited anonymity could be assured; that is, it was entirely possible that members of a given team could recognize comments made within the discussion forums and attribute them to specific team members.

Table 9.1 Pseudoexamples of posts coded to CoI cognitive presence indicators

Source	Indicator	Example post
Garrison et al. (2001)	Triggering	What waste does road work produce?
	Exploration	I think we should include amphibians too…? I found a list of contaminants that are released by that industry (citation 1)
	Integration	There's evidence from (citation 1 & citation 2) that back up my idea about how the pollutants move
	Resolution	Here is a first draft of the report—everyone please comment!
	Other	That's an awesome reference!
Noble (2017)	Organization	We are meeting in the breakout room at noon tomorrow Please add information about your chosen chemicals to this table Who can take on looking up steel mill processes?

Each post was coded to a single category because the CoI cognitive presence indicators are meant to be mutually exclusive (Garrison et al., 2001). If a post contained indicators that fit more than one category, I adopted the *coding-up* approach used by Garrison et al. (2001). If a post contained clear evidence of both exploration and integration (e.g., brainstorming supported by several citations), the post would be coded as integration because the student showed signs of higher-level thinking. For example, in the post

> Do you think it is worth looking more into pulp mill contaminants? So far, I have found that not only is chlorine an issue, the same source suggests that they are a major contributor of dioxins and endocrine-disrupting chemicals. (citation)

the first statement is a trigger but the statement that follows is an exploration with a supporting citation; therefore, the post would be coded as exploration. As the sole coder, I followed Rourke et al. (2001) and checked my consistency by recoding the 492 posts from one distance cohort and one blended cohort twice after completing the initial coding. My coding consistency was robust with only 12 differences in the first recoding and none in the second.

Developing Thread Maps

Once the coding was completed, I examined the discussion forums qualitatively. I created thread maps (Levin et al., 1990) to examine how the students were approaching their asynchronous discussions about the PBL case problems, particularly the timing and organization of an online discussion in conjunction with the CoI coding (see Fig. 9.2). Online discussion posts are displayed under headings, and the use of a new heading signals a new thread. Each post is represented as a line on the thread map that included the post title, author's name, date and time of posting, and post length (i.e., number of lines of text or number of pages in an attachment). Posts that started a new thread were at the left margin on the diagram; any replies

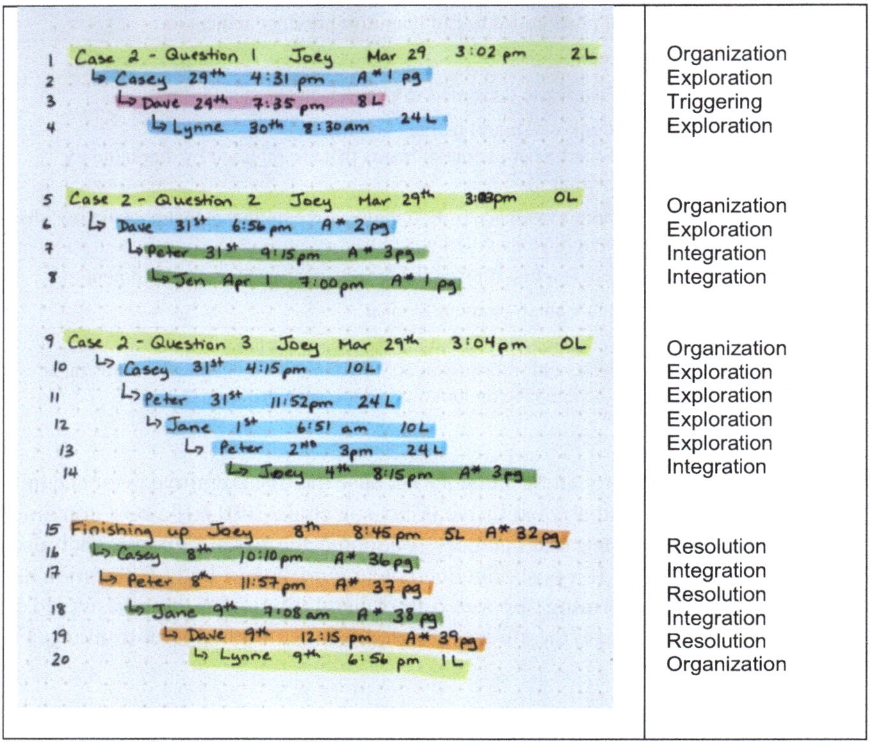

Note. Colour coding: yellow = organization; pink = triggering; blue = exploration; green – integration; orange = resolution; white = other.

Fig. 9.2 Thread map pseudoexample demonstrating colour-coding of CoI indicators. *Note* Colour coding: yellow = organization; pink = triggering; blue = exploration; green – integration; orange = resolution; white = other

to a thread were indented below the primary post and connected to it by an arrow. Different colours indicate each CoI cognitive indicator. I analyzed the thread maps for group interactions including the use of the problem's scaffolds.

9.3 Results and Discussion

In this section, I present and discuss the results of the quantitative and qualitative analyses. I highlight similarities and differences between the distance and blended groups with respect to academic outcomes, content analysis based on the CoI indicators, and thread maps.

9.3.1 *Academic Outcomes*

Comparison of the mean examination scores for the distance ($M = 80.59$, $SD = 12.80$) and blended ($M = 81.10$, $SD = 12.01$) students indicated that both groups demonstrated a similar level of achievement with a similar spread of grades. An independent samples *t* test confirmed that there was no significant difference in the mean scores of the distance and blended groups ($t(250) = 0.32, p = 0.747$). Similarly, an independent samples *t* test confirmed that there were no significant differences in the mean scores of the distance ($M = 22.35$, $SD = 0.856$) and blended ($M = 22.31$, $SD = 0.447$) groups on the three PBL case problems ($t(32) = 0.94$, $p > 0.05$).

9.3.1.1 CoI Indicators

For the four distance cohorts, there were a total of 8,313 posts distributed across 797 threads made by 15 teams. For the four blended cohorts, there were a total of 3,671 posts distributed across 919 threads by 19 teams. The frequencies of codes for the distance and blended cohorts are shown in Table 9.2.

In order to explore these results more closely and remove any obfuscation due to the high frequency of posts coded other and organization, I focused on the CoI indicators (see Table 9.3). It appears that there were differences between the two formats; however, those apparent differences disappeared when the organization and other posts were removed.

Similarly, to explore the usage of my new organization category, I focused on the results for posts coded other and organization and compared their frequency between cohorts. Separating posts in this way allowed for the examination of patterns of use (see Table 9.4). Again, percentages of each type of post were similar for both formats although the total number of posts was five times higher for the distance teams.

Table 9.2 Frequencies of codes for distance and blended formats

Code	Distance (60 discussions)		Blended (76 discussions)	
	n	%	*n*	%
Triggering	33	0	48	1
Exploration	1,861	22	1,487	41
Integration	1,535	18	1,126	31
Resolution	269	3	119	3
Other	705	8	119	3
Organization	3,910	47	772	21
Total	8,313	100	3,671	100

Table 9.3 Frequencies of CoI indicators for distance and blended formats, posts coded other and organization removed

Indicator	Distance (60 Discussions)		Blended (76 Discussions)	
	n	%	*n*	%
Triggering	33	1	48	2
Exploration	1,861	50	1,487	53
Integration	1,535	42	1,126	41
Resolution	269	7	119	4
Total	3,698	100	2,780	100

Table 9.4 Frequency of other and organization posts for distance and blended formats

Code	Distance (60 discussions)		Blended (76 discussions)	
	n	%	*n*	%
Other	705	15	119	13
Organization	3,910	85	772	87
Total	4,615	100	891	100

9.3.1.2 Thread Maps

Next, I analyzed the thread maps for patterns with attention to the CoI cognitive indicators and my newly identified organization indicator. I also searched for the use of the PBL scaffolds provided within the case problems. When these indicators were transferred to the thread maps, the patterns highlighted two major approaches that students were taking to the online discussions, which I labelled as organic or structured. Organic discussions had no evident overarching structure and included limited or no use of instructional scaffolds. In organic discussions (Fig. 9.3), team members started threads on an as-needed basis; and it was not uncommon for a thread to contain only one or two posts. Structured discussions had an overarching structure based on the provided instructional scaffolds. In structured discussions (Fig. 9.4), the student moderators used the instructional scaffolds to start separate threads to organize the discussion into sections. Only the moderator started threads in a structured discussion and all team members contributed information as replies to these initial posts, resulting in threads that contained multiple related posts. Although organic and structured styles were visible in both distance and blended team discussions, 54% of the blended teams' discussions were organic and 46% were structured while 35% of the distance teams' discussions were organic and 65% were structured.

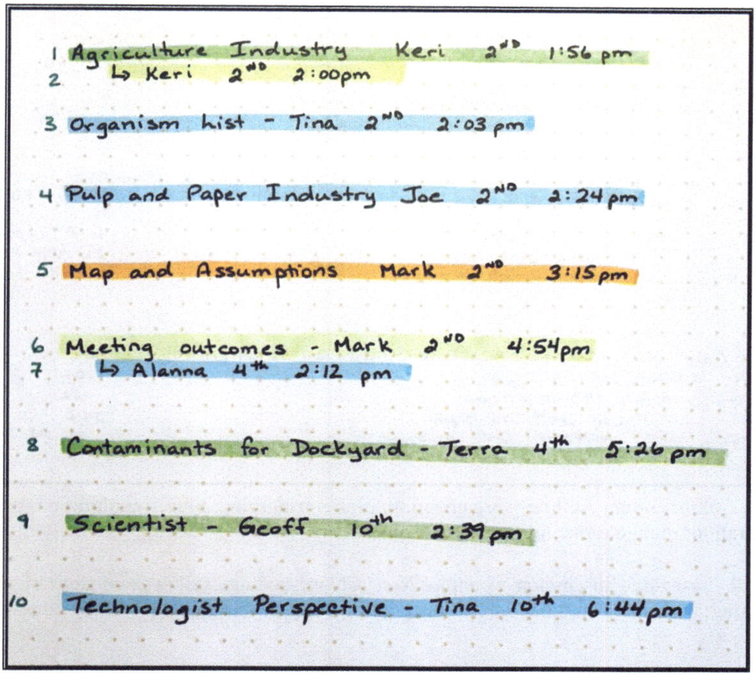

Note. Colour coding: yellow = organization; pink = triggering; blue = exploration; green – integration; orange = resolution; white = other.

Fig. 9.3 Organic discussion example. *Note* Colour coding: yellow = organization; pink = triggering; blue = exploration; green – integration; orange = resolution; white = other

9.4 Important Outcomes/Contributions

Here I discuss the similarity of the academic outcomes for distance and blended teams. I also explain the patterns that emerged once the CoI indicators were applied to the discussion forums including why the organization indicator was added to the cognitive presence coding, provide quantitative comparisons of cognitive presence indicators between distance and blended students as well as comparisons to other studies, and discuss what the analysis of the thread maps suggested about group interactions and the impact of blended versus distance cohorts.

9.4.1 Academic Outcomes

The academic outcomes for the distance and blended groups were not appreciably different for either the four PBL case problems or the PBL-based scenario question

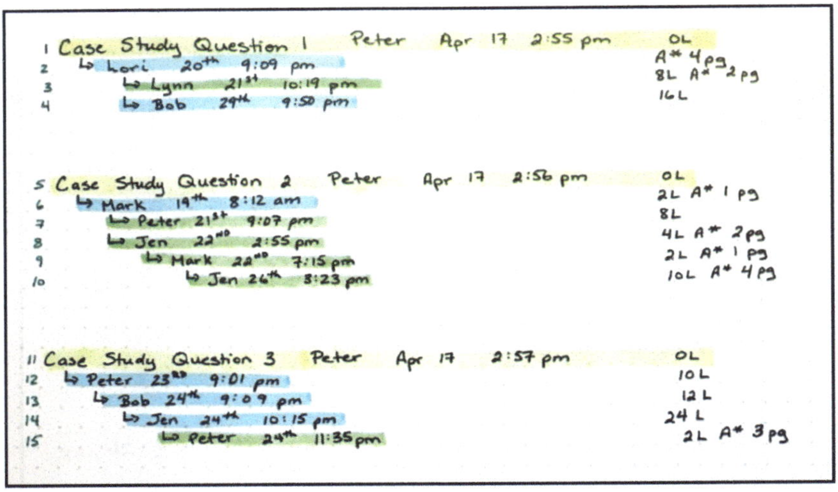

Note. Colour coding: yellow = organization; pink = triggering; blue = exploration; green – integration; orange = resolution; white = other.

Fig. 9.4 Structured discussion example. *Note* Colour coding: yellow = organization; pink = triggering; blue = exploration; green – integration; orange = resolution; white = other

on the final examination. This result supports the conclusion that students were achieving very similar outcomes whether their primary communication with their team was occurring online or face-to-face. In future, it may be interesting to see if scores on individual case problems influenced the choice of organic versus structured discussions; that is, did high scores contribute to a team maintaining its choice of discussion style or a low score lead to a team deciding to change its discussion style.

9.4.2 New Indicator: Organization

The organization indicator was added to the coding for cognitive presence because the other category was capturing posts that seemed to require actions such as planning, creation of templates, and timelines. These posts did not fit into the social (i.e., open communication, group cohesion, or affective expression) or teaching (i.e., design and organization, facilitating discourse or direct instruction) presence indicator parts of the CoI framework. At first glance, the design and organization indicator of teaching presence, which describes planning and design of the structure, process, interaction and evaluation aspects of the online course (Anderson et al., 2001; Garrison & Arbaugh, 2007), appeared to be a possible fit for these posts. However, because that category captures instructor rather than student activities, adding a new category captured peer-to-peer interactions that supported the cognitive functioning of teams (see the area of Fig. 9.1 that represents the overlap between cognitive presence and teaching presence). As noted by Vaughan et al., "All participants in a collaborative learning environment [like a PBL team] must assume various degrees of teaching responsibilities depending on the specific content, developmental level, and ability" (2013, p. 14). Inclusion of organization with the cognitive presence indicators rather than the teaching or social presence indicators is supported by several observations. The cognitive behaviours observed in organization posts included volunteering to research a topic, setting deadlines, dividing a large task into smaller ones, using the problem's scaffolds to organize the discussion threads, and setting up meetings. The similarity of these behaviours to the instructor's role in organizing the curriculum and activities of the course reflects the peer teaching that happens in collaborative learning environments like PBL teams.

Such peer-to-peer teaching presence is also evidence that students were trying "to assume increasing responsibility and control of their learning" (Vaughan et al., 2013, p. 13), which is a goal of PBL in general (Guilar & Loring, 2008; Hmelo-Silver et al., 2007) and the LTRM in particular (RRU, n.d.). Thus, the placement of the organization category at the intersection of cognitive and teaching presence (Fig. 9.1) may provide indicators for self/group-regulated learning and give insight into the team process of organizing thoughts and approaches to problem solving. Although the present study did not use this lens to examine the data, future work may examine this indicator from the self/group-regulated learning perspective to provide further insight to the metacognitive processes occurring within online PBL teams.

9.4.3 Comparison of Discussions in Distance and Blended Formats

The percentage of exploration and integration posts appears smaller for the distance group than for the blended group, likely a reflection of the blended teams organizing themselves and collaborating offline while distance team discussion forums had a higher proportion of posts coded as organization (Table 9.3). This interpretation is supported by the percentage frequencies of the exploration and integration posts coming back into alignment when the organization and other posts are removed (Table 9.4). Integration posts for both blended and distance teams usually contained information with proper citations and fully formatted paragraphs that were ready to be integrated into the final report. In addition, distance teams more frequently assembled their final report documents on the discussion forum, which resulted in more integration posts than observed for blended teams as the sections of the report were built up. The thread maps and content analysis of the posts by both groups support this interpretation of frequency counts for each of the cognitive presence indicators.

Similar to previous work by Meyer (2003) and Garrison et al. (2001) that examined online non-PBL discussions, the exploration and integration categories were the most frequently used CoI cognitive categories for both distance and blended teams when social or other posts are removed from consideration. Interestingly, Meyer (2003) and Garrison et al. (2001) reported much lower percentages of integration posts (22% & 19%, respectively) in their studies of non-PBL discussions than the 41% of distance posts and 42% of blended posts observed here. This suggests that the need to produce a report to address each PBL problem drove students to progress beyond the exploration stage of discussion and provided an opportunity for students to integrate their own ideas as well as ideas from diverse source materials.

The other category consisted of posts that did not fit elsewhere even after the addition of the organization indicator. These posts were often social interactions such as off-topic discussions, jokes, or other conversations. Xin and Feenberg (2007) pointed out that "the more deeply the participants interact, the more successful they will be in advancing the agenda of the course and achieving individual conceptual change" (p. 422). Both the distance and blended students had formed tight-knit learning communities at this point in their program as is typical for RRU student cohorts (Guilar & Loring, 2008), and these close relationships likely helped to support the teams' learning. The distance teams posted nearly ten times as many other messages per case than did the blended teams (see Table 9.3), likely because for distance teams the discussion forums were their primary mode of communication and the distance team used posts to reinforce their social dynamics (Garrison & Arbaugh, 2007). In many ways, these other posts represented the online equivalent of chatting with classmates before or after class. Because the blended teams could reinforce their social interactions in face-to-face meetings, they were less likely to make social posts in the discussion forum.

Blended teams were less likely to post in general but also made fewer organization/other posts than distance teams, indicating that substantial activity was occurring offline for the blended teams. Although the ability to meet in person potentially gave the blended teams an advantage, especially in terms of being able to quickly reach decisions, this potential advantage did not seem to translate into better overall results on the PBL case reports where both distance and blended teams achieved very similar academic outcomes.

9.4.4 Discussion Style

In this section I discuss the use of organic and structured discussion styles and the association with teams' cognitive presence when the CoI cognitive presence indicators were overlaid on the thread maps. As noted earlier, while approximately two-thirds of the distance team discussions used structured discussion styles, only half of the blended team discussions did so. The choice of discussion style could be associated with the frequency of posts that contained cognitive presence indicators because structured discussions tended to contain more integration posts where students were collaboratively constructing knowledge. The greater use of the structured approach by distance teams compared to blended teams may reflect the distance students' need to organize the information contained in the much greater number of posts overall.

When teams used the organic discussion style, threads were started on an as-needed basis by the message author. For blended teams, this style often resulted in very short threads because students started a new thread each time they posted a message and very little, if any, discussion occurred; these posts were mostly exploratory with some integration posts. This gave the discussion forum the appearance of a repository meant to allow student moderators to collect the teams' ideas for incorporation into the final report rather than a collaborative discussion. Although distance teams that used the organic style had longer threads and featured discussion of ideas with teammates (resulting in more integration of ideas) that would be expected when the discussion forum is the primary mean of communication, the threads were still shorter than those of distance teams that used a structured approach.

The structured discussion style allowed teams to group related messages and encouraged the generation of integration posts as students built up ideas collaboratively, which greatly increased the cognitive presence as shown by CoI indicators. Student moderators used the scaffolds provided in the PBL case problem narrative as the headers for organizational posts before other team members began posting. These scaffolds guided students by providing cues or prompts to help them limit the scope of the problem and supported students' confidence in their problem-solving ability (Jonassen, 2000) by reducing the ill-structured nature of the problem to a more manageable level. This confidence was especially important in the first two case problems because students were still learning how to be successful in PBL and in the course overall.

Most teams used a discussion style consistently across all four case problems. In situations where teams changed their discussion approach, there did not seem to be any pattern to the change. Some teams started by using a structured style in early case problems and moved to an organic style for later case problems; other teams started with an organic style and moved to the use of scaffolds in a structured style; and still other teams alternated between both styles. In future, I would like to examine what drives the choice between organic and structured discussions and the reasons for changes between strategies. For example, is this completely a choice driven by the moderator? Does academic success or lack thereof in a previous PBL problem influence the choice of style for the next problem?

9.5 Conclusions and Considerations

The last 20 years have seen a major shift in PBL research with the recognition that PBL encompasses many variables including problem solving, teamwork, research, and communications. This recognition has led to investigations of more subtle variables that may contribute to PBL's effects, including the format in which PBL is presented. In this study, I considered how PBL assignments might be undertaken by fully asynchronous or blended teams by examining the patterns that emerged when the CofI indicators were applied to assess the activity in the discussion forums.

When examining the online activity of groups engaged in PBL, the organization indicator for cognitive presence captures student activities like peer teaching that are similar in function to the instructor's role in organizing the curriculum and activities of the course that are captured in teaching presence indicators. The organization indicator illuminates the style that the team is using for the discussion and provides a useful starting point for discussion between the instructor and the team to address difficulties that stem from their approach either to the problem or their moderation skills. This has allowed me to coach these problem-solving skills and help students understand how planning steps (e.g., using templates to scaffold filling information gaps) can be useful to their team process.

Moving forward, one option that I am exploring is coaching students to use the organizational discussion style. Although it could be argued that such coaching may limit students' discovery of the best strategy to demonstrate their cognitive presence, this additional teaching presence support may help to limit struggles around the group process, leaving more capacity for students to wrestle with content and identifying knowledge gaps. It would be interesting to see if such coaching would result in stronger academic outcomes or more streamlined team processes.

My experience is that online discussions are a useful component of both the distance and blended classes. Online discussions provide a means for students to develop critical skills as they write about what they are learning. By encouraging students to write posts, they develop skills in writing, linking ideas, and critically evaluating source materials. I can provide scaffolding and modelling of appropriate behaviour as students develop feedback, questioning, and problem-solving skills. Online discussions also provide an increased opportunity for reflective thinking, making online PBL a different experience than face-to-face PBL. Additionally, because of the requirement to submit a summary report of the team's solution, students need to discuss and collaboratively assemble their ideas and understandings, causing them to move beyond just brainstorming.

One of PBL's strengths is how the problems can be used to show students (a) the complexity of the interconnections between ideas and disciplines in a way that rewards creativity and (b) the ability to understand a situation from a variety of perspectives. This is particularly true for fields like environmental science that by their interdisciplinary scope draw on a variety of tools, concepts, and skills. In the context of this study, PBL enabled the learning of ecotoxicology to be firmly anchored within the reality of life in the world, thus encouraging these students to be engaged in self-reflection and practical applications of their learning.

PBL can promote dialogue among peers, leading to a deeper understanding of applied concepts, support teamwork and the formation of learning communities, and encourage students to take responsibility for their learning and the learning of their team (Guilar & Loring, 2008; Hmelo-Silver, 2004, 2007); online PBL provides all these benefits with the addition of a written record. Since the written record of the online PBL discussion can be revisited, added to, and discussed at any time in a way that a face-to-face meeting cannot, online PBL provides increased time for reflection while the problem solving is in progress. The addition of the organization indicator to the CoI cognitive presence indicators provides a valuable window into peer-to-peer interactions, including peer teaching, allowing for the critique and coaching of many aspects of the PBL process, concepts, and approach in a more nuanced way that will hopefully inspire better teaching and learning.

Acknowledgements I would like to thank Dr. Tim Pelton for his support throughout this project, Royal Roads University for funding support, and Dr. Norman Vaughn for his insightful comments on this research.

References

Anderson, T., Rourke, L., Garrison, D. R., & Archer, W. (2001). Assessing teaching presence in a computer conferencing context. *Journal of Asynchronous Learning Networks, 5*(2), 1–17. https://onlinelearningconsortium.org/read/journals/journal-asynchronous-learning-networks/

Arbaugh, J. B. (2007). An empirical verification of the community of inquiry framework. *Journal of Asynchronous Learning Networks, 11*(1), 73–85. https://onlinelearningconsortium.org/read/journals/journal-asynchronous-learning-networks/

Barrows, H. S. (1996). Problem-based learning in medicine and beyond: A brief overview. *New Directions for Teaching & Learning, 1996*(68), 3–12. https://doi.org/10.1002/tl.37219966804

Barrows, H. S. (2002). Is it truly possible to have such a thing as dPBL? *Distance Education, 23*(1), 119–122. https://doi.org/10.1080/01587910220124026

Dochy, F., Segers, M., van den Bossche, P., & Gijbels, D. (2003). Effects of problem-based learning: A meta-analysis. *Learning & Instruction, 13*(5), 533–568. https://doi.org/10.1016/S0959-4752(02)00025-7

Garrison, D. R. (2017). *E-learning in the 21st century: A community of inquiry framework for research and practice* (3rd ed.). Routledge.

Garrison, D. R. (2007). Online community of inquiry review: Social, cognitive and teaching presence issues. *Journal of Asynchronous Learning Networks, 11*(1), 61–72. https://onlinelearningconsortium.org/read/journals/journal-asynchronous-learning-networks/

Garrison, D. R., Anderson, T., & Archer, W. (2001). Critical thinking, cognitive presence, and computer conferencing in distance education. *American Journal of Distance Education, 15*(1), 7–23. https://doi.org/10.1080/08923640109527071

Garrison, D. R., & Arbaugh, J. B. (2007). Researching the community of inquiry framework: Review, issues and future directions. *The Internet & Higher Education, 10*(3), 157–172. https://doi.org/10.1016/j.iheduc.2007.04.001

Garrison, D. R., Cleveland-Innes, M., & Fung, T. (2004). Student role adjustment in online communities of inquiry: Model and instrument validation. *Journal of Asynchronous Learning Networks, 8*(2), 61–74. https://onlinelearningconsortium.org/read/journals/journal-asynchronous-learning-networks/

Garrison, D. R., Cleveland-Innes, M., & Vaughan, N. D. (n.d.). *Community of inquiry framework.* https://coi.athabascau.ca

Gijbels, D., Dochy, F., van den Bossche, P., & Segers, M. (2005). Effects of problem-based learning: A meta-analysis from the angle of assessment. *Review of Educational Research, 75*(1), 27–61. https://doi.org/10.3102/00346543075001027

Guilar, J. D., & Loring, A. (2008). Dialogue and community in online learning: Lessons from Royal Roads University. *International Journal of E-Learning & Distance Education / Revue internationale du e-learning et la formation à distance, 22*(3), 19–40. https://www.ijede.ca/index.php/jde/article/view/52

Heckman, R., & Annabi, H. (2005). A content analytic comparison of learning processes in online and face-to-face case study discussions. *Journal of Computer-Mediated Communication, 10*(2), JCMC10210. https://doi.org/10.1111/j.1083-6101.2005.tb00244.x

Hmelo-Silver, C. E. (2004). Problem-based learning: What and how do students learn? *Educational Psychology Review, 16*(3), 235–266. https://doi.org/10.1023/B:EDPR.0000034022.16470.f3

Hmelo-Silver, C. E., Duncan, R. G., & Chinn, C. A. (2007). Scaffolding and achievement in problem-based and inquiry learning: A response to Kirschner, Sweller, and Clark (2006). *Educational Psychologist, 42*(2), 99–107. https://doi.org/10.1080/00461520701263368

Jonassen, D. H. (2000). Toward a design theory of problem solving. *Educational Technology, Research & Development, 48*(4), 63–85. https://doi.org/10.1007/BF02300500

Koschmann, T. D., Myers, A. C., Feltovich, P. J., & Barrows, H. S. (1994). Using technology to assist in realizing effective learning and instruction: A principled approach to the use of computers in collaborative learning. *Journal of the Learning Sciences, 3*(3), 227–264. https://doi.org/10.1207/s15327809jls0303_2

Levin, J. A., Kim, H., & Riel, M. M. (1990). Analyzing instructional interactions on electronic message networks. In L. M. Harasim (Ed.), *Online education: Perspectives on a new environment* (pp. 185–213). Praeger.

Meyer, K. A. (2003). Face-to-face versus threaded discussions: The role of time and higher-order thinking. *Journal of Asynchronous Learning Networks, 7*(3), 55–65. https://onlinelearningconsortium.org/read/journals/journal-asynchronous-learning-networks/

Noble, M.-A. (2017). *Characterization of critical thinking indicators in problem-based learning online discussions of blended and distance undergraduate environmental science students using the community of inquiry model* [Doctoral dissertation, University of Victoria]. https://dspace.library.uvic.ca//handle/1828/8516

Pena-Shaff, J. B., & Nicholls, C. (2004). Analyzing student interactions and meaning construction in computer bulletin board discussions. *Computers & Education, 42*(3), 243–265. https://doi.org/10.1016/j.compedu.2003.08.003

Rourke, L., Anderson, T., Garrison, D. R., & Archer, W. (2001). Methodological issues in the content analysis of computer conference transcripts. *International Journal of Artificial Intelligence in Education, 12*, 8–22.

Royal Roads University. (n.d.). *Learning, teaching, and research model.* https://www.royalroads.ca/about/academics/learning-teaching-research-model

Schmidt, H. G., Rotgans, J. I., & Yew, E. H. J. (2011). The process of problem-based learning: What works and why. *Medical Education, 45*(8), 792–806. https://doi.org/10.1111/j.1365-2923.2011.04035.x

Strobel, J., & van Barneveld, A. (2009). When is PBL more effective? A meta-synthesis of meta-analyses comparing PBL to conventional classrooms. *Interdisciplinary Journal of Problem-Based Learning, 3*(1), Article 4. https://doi.org/10.7771/1541-5015.1046

Vaughan, N. D., Cleveland-Innes, M., & Garrison, R. D. (2013). *Teaching in blended learning environments: Creating and sustaining communities of inquiry.* AU Press. https://doi.org/10.15215/aupress/9781927356470.01

Walker, A., & Leary, H. (2009). A problem based learning meta analysis: Differences across problem types, implementation types, disciplines, and assessment levels. *Interdisciplinary Journal of Problem-Based Learning, 3*(1), Article 3. https://doi.org/10.7771/1541-5015.1061

Xin, C., & Feenberg, A. (2007). Pedagogy in cyberspace: The dynamics of online discourse. *E-Learning & Digital Media, 4*(4), 415–432. https://doi.org/10.2304/elea.2007.4.4.415

Michael-Anne Noble BSc (University of Victoria), MSc (University of Victoria), PhD (University of Victoria), is a professor in the School of Environment and Sustainability at Royal Roads University, teaching applied microbiology and biochemistry since 1996. Her current research interests include problem-based learning in online environments, undergraduate science curriculum design and children's representations of scientists. She developed and works extensively with an elementary science outreach program to allow elementary school children the opportunity to experience hands-on science both in the laboratory as well as in the forest and creeks on the Royal Roads grounds.

Chapter 10
Science Teachers in an Immersive Museum Education Program: A Social Theory of Learning Perspective

Anik Meunier and Charlène Bélanger

10.1 Introduction

In the early 2000s, a major reform in initial and inservice teacher education was implemented in Québec, based on a new conception of the teacher as teaching professional. In this context, competencies set out in the *Référentiel de compétences professionnelles de la profession enseignante* [Reference Framework for the Teaching Profession based on Professional Competencies] (Ministry of Education, 2020) included the culture of teachers and their ability to use cultural objects in their professional practice. Competency 1 was "To act as a professional inheritor, critic and interpreter of knowledge or culture when teaching students" (Martinet et al., 2001, p. 54).

The role of the teacher is, therefore, recognized from the outset as a "cultural transmitter" to students who are encouraged to take advantage of the many cultural institutions and resources that exist outside the school. Teachers can spark student interest in knowledge and culture and develop the habit of using these resources (Zakhartchouk, 1999). Although this competency focuses on teacher-as-cultural-transmitter as part of their professional role, it is rarely enacted using cultural institutions outside the school (Bitgood, 1989; DeWitt & Storksdieck, 2008; Griffin & Symington, 1997; Meunier & Bélanger, 2014). Here we present the use of a museum as an external cultural resource because we are interested in observing the science teaching and learning opportunities to which museums can contribute.

Research on teaching practices related to museums—informal learning contexts including zoos, aquaria, and conservation areas—has shown that most teachers who

A. Meunier (✉)
University of Quebec in Montreal (UQAM), Montreal, QC, Canada
e-mail: meunier.anik@uqam.ca

C. Bélanger
Planetarium de Montréal, Montreal, QC, Canada
e-mail: charlene.belanger@montreal.ca

C. D. Tippett and T. M. Milford (eds.), *Seeing Science Through the Eyes of Canadian Teachers and Learners*, Contemporary Trends and Issues in Science Education 54,
https://doi.org/10.1007/978-3-031-88785-7_10

use museums consider them as an educational tool to support teaching. They want museums to develop exhibitions and educational activities closely related to the prescribed curriculum. In science specifically, research has shown that a majority of teachers, especially at the primary level, expect museums to fill curricular gaps (Cohen, 2003). Given these practices, we are removed from the idea of a museum visit as a cultural activity. Museums are used as a tool for schools and are not recognized for their full learning potential for students as future lifelong learners (Bélanger & Meunier, 2012) in informal educational settings (Meunier, 2018).

Teachers in general are themselves rarely museum visitors (Gajardo, 2005). In a multicase study, Gajardo compared the teaching practices related to museums between non-visiting and regular museum-visiting teachers. She demonstrated that teachers who do not generally visit museums have representations of a museum as an educational tool to serve their teaching needs. Conversely, teachers who reported regularly being museum visitors had a personal relationship with these institutions and shared with their students their interest in the museum as a cultural space and rich source of learning. Her study demonstrated that in order to satisfy their role as cultural transmitters teachers must develop a personal relationship with museums by familiarizing themselves with these spaces and by becoming museum visitors themselves. Only after they have developed this relationship with museums can they share that interest and passion with their students. A recent study by *Le Groupe de recherche sur l'éducation et les musées* [Education and Museum Research Group; GREM] on the topic of teachers who do not frequent museums showed the same results (Meunier et al., 2021).

Teacher education programs can help teachers discover the resources of museums and other cultural institutions. A study of Québec university education programs, however, showed that none of the programs included courses that cover the use of cultural and heritage resources for educational purposes (Bélanger & Meunier, 2012; GREM, 2000). This aspect of teaching is left to the initiative of professors who may or may not introduce the topic of using outside school resources in courses related to specific school subjects (e.g., social studies or science).

Recognizing their limited use by teachers, many museums have developed teacher education programs. Bélanger and Meunier (2012) developed a typology of education programs: subject-related, teaching-related, related to the educational use of museums, and museum education. While the first three program types place museums at the service of schools and strictly academic learning, museum education programs seem the most promising in the context that concerns us, namely, adapting schools for students in the twenty-first century. In a learning society, all members are invited to the vast undertaking of social learning; school does not hold a monopoly over education anymore (Daoust et al., 1974; Hutchins, 1969). School is but an integral part of a larger infrastructure encompassing non-formal education settings, which have seen the development of diversified offerings to satisfy the needs of lifelong learning. The tremendous work taking place within these settings must be considered (Falk & Dierking, 2018; Falk et al., 2006).

The aim of museum education programs is not to equip teachers in terms of their teaching or to help them establish links between museum content and elements of school curricula; instead, it is to familiarize teachers with the world of museums and museum practices so that teachers are more confident and independent in that environment. They aim to develop a personal relationship between teachers and museums in the hope that teachers will share this interest with their students and, thus, fully assume their role as cultural transmitters.

10.1.1 The Museum Pedagogy Program

The museum education program that is the subject of this chapter was initiated and coordinated by the *Centre muséopédagogique* [Museum Education Centre; MEC] of the *Musée de la civilisation* [Museum of Civilization] in Québec City. The MEC's mission is to provide inservice teacher education that supports teachers to become cultural transmitters. The MEC allows member–teachers (i.e., teachers who pay an annual fee) to use its human and material resources to develop personalized learning situations for their students. In addition, groups of teachers are occasionally invited to participate in special museum projects, such as developing interdisciplinary discovery spaces in the museum's exhibition halls or interpretive tours for school-age children.

Our chapter examines a specific 24-month collaborative initiative to design a science exhibition for secondary school students between the MEC and a group of secondary school science teachers. The purpose of the exhibition was to familiarize students with specific science content related to school curricula and to introduce various science careers. For example, the teachers wanted to make teaching and learning situations more meaningful by providing concrete examples where science and technology were applied, particularly in the design of high-performance sports equipment. From the point of view of the museum staff, the presence of science teachers during the design process was justified by the need to test ideas with pedagogical experts. In addition, the MEC team that developed this program considered the immersion experience for teachers as unconventional professional development. Indeed, in contrast to typical preservice teacher education that aims to equip teachers in terms of their classroom practice, this museum program aimed to familiarize teachers with the museum environment and way of doing things. The goal was that the teachers would develop a personal relationship with the museum and then share this interest with their students.

As researchers in education and museology, we were interested in the unique character of this continuing education program. We observed the teachers throughout the immersion process in the museum's activities in order to describe and better understand how this experience changed their teaching practice and their relationship with museums.

10.1.2 Theoretical Framework

Teacher education activities today are most often geared toward the development of a reflective attitude that allows for continuous and autonomous change in teaching practice (El Islami et al., 2022; Popova et al., 2022). To foster reflection, communities of practice can bring about change in practice and professional development (Grossman et al., 2001; Yamagata-Lynch, 2001). Communities of practice are groups of individuals working in one professional context, maintaining close relationships and sharing specific practices (Lave & Wenger, 1991; Wenger, 1998). To understand the learning that occurs in this kind of context, a social theory of learning (Wenger, 1998, 2018) can be applied.

Learning is described as a system of four dimensions that are closely related and mutually determining. The first dimension includes *meaning*, that is, the ability of individuals to derive different meanings from their experience. The second dimension is concerned with the changes arising in the collective use of cultural resources, perspectives, and frameworks that support mutual engagement in *practice*. The third dimension concerns the changes that relate to belonging to a *community* and involves relationships among individuals, the emerging positioning of individuals, and collaborations. The fourth dimension describes the changes in the *identity* of members of a community of practice; it encompasses how participation in collective activities transforms individuals and creates personal stories of identity development within communities. The four dimensions are intimately linked to each other, influencing each other and overlapping.

Wenger (1998) observed that communities of practice are not defined by watertight borders. Instead, exchanges between communities are made possible through boundary encounters (e.g., meetings, conversations, visits) between members of distinct communities or through the use of boundary objects (e.g., guidelines, maps, exhibitions). The social theory of learning positions boundary encounters as a powerful engine for learning. Contacts between boundary brokers (e.g., members from the communities involved in the encounter) and the mobilization of boundary objects can destabilize or strengthen learning of the community members (c.f. Wenger-Trayner et al., 2023).

In our study at the boundary between science museums and schools, the boundary object was the science exhibition to be planned that would display artefacts, specimens, and natural phenomena in three-dimensional spaces to present a certain conception of the world specific to the scientific community. The exhibition allowed the boundary brokers (i.e., the schoolteachers and the museum professionals) to interact and collaborate in a new space arising from the encounter between their two distinct communities of practice.

10.2 Method

We were concerned with exploring the collective learning experience of science teachers who were immersed in an unconventional continuing education program in a museum that explores the human experience (Musée de la civilisation, 2024). We developed a qualitative–interpretive study focusing on an inservice program implemented by MEC staff for a group of six secondary school science teachers. Only five of these teachers participated in the initial interview, and only four teachers participated in the final interview; in this chapter we focus on these four teachers. The interpretive approach allowed us to understand the participants' learning experience while considering their point of view.

The teachers in the program were recruited from school boards in Québec City and had volunteered to participate after information was shared with the boards. During the program, the teachers were exempt from their school duties so that they could spend a full day at the museum every 4 or 6 weeks. The program's first year of activities included discovering the museum, its internal operations and its physical spaces—exhibition halls, administrative spaces, and collections. The teachers had the opportunity to meet museum professionals and become familiar with the specific work of exhibition design teams. In the second year, they were involved in the design of an exhibition for secondary school students on the theme of *la science dans les sports d'élite* [the science behind high-performance sport].

Each teacher was responsible for a specific topic about elite sports (e.g., alpine ski equipment, wheelchairs for racing, 100 m sprint shoes, swimming clothing) and developing concepts about that topic for a part of the exhibit area. Meeting five or six times during the school year, the teachers discussed how to best convert their topic into an exhibition. The rest of the design work was conducted by the museum staff. The exhibition *Science en direct / Science Live* opened to the public in June 2011 and ran until August 2012. Figures 10.1 and 10.2 show the exhibition entry and a student taking part in the exhibit Cultiver la culture de la science / *Cultivating the culture of science*.

We conducted two sets of semi-structured interviews with the teachers, the first at the end of the 24-month program and the second one year post program. The interview guide for the first interview was based on objectives formulated by MEC staff and dealt with the effect of study participation on teacher practices with regard to museums; classroom teaching practices; the meaning given to museums and to school visits, including pedagogical goals/objectives; the role museums can play in science education; the teachers' feeling of self-competence within the framework of the study; and what motivated them to participate in the study. For the second interview, questions were personalized based upon participating teachers' responses during the first interview. For example, if in the first interview someone told us about their intention to visit the museum with their students, we asked them in the second interview about that intention and whether they actually had visited the museum.

Fig. 10.1 Entry to the Science en direct (Science Live) exhibition

Fig. 10.2 Exhibit *Cultivating the Culture of Science* (Cultiver la culture de la science)

10.3 Analysis and Results

We used the framework of the social theory of learning to analyze the responses of the four teachers who participated in both interviews. Through the lens of the four dimensions of meaning, practice, community, and identity (Wenger, 2018), we paid particular attention to the changes undergone by the teachers. We used both pure inductive (Creswell & Creswell, 2022) and deliberative inductive approaches (Savoie-Zajc, 2011) to characterize the participating teachers' experiences. In the following sections, we describe the characteristics of each teacher (designated by a pseudonym), discuss their reasons for participating in the study, then we present their experiences regarding each component of Wenger's social theory of learning.

10.3.1 Portraits of the Four Teachers

Xavier was a physics teacher with approximately 10 years of teaching experience. He taught in a small disadvantaged secondary school in the region of Québec City. His school had three science teachers; he was the only Secondary IV and V (Grades 10 & 11) physics teacher. Each year, he organized a museum trip for Secondary IV students in collaboration with other teachers at the school (e.g., ethics, history, and geography teachers). Xavier was convinced of the positive impact of the museum visit on his students.

Hélène had less than 5 years of teaching experience. She taught Secondary I and II (Grades 7 & 8) science at a school located on the outskirts of Québec City. She described herself as a museum visitor but had not visited a museum with her students due to time and budget limitations. For her, school trips were above all an opportunity for students to discover different trades and professions and be inspired by new career choices. Rémi had more than 20 years of teaching experience in the same school board; he began the program with great confidence in himself, his expertise, and his ability to know what appealed to secondary students. He taught Secondary IV in a school that was outside Québec City. He described the school's population as economically disadvantaged. Since the cultural resources of his region were limited compared to large urban centres, he considered it his responsibility to organize regular school trips to large museums.

Martine was a Secondary IV and V science teacher who had about 10 years of teaching experience. Her school offered gifted students the option to complete the International Baccalaureate program. Martine was fortunate to have a collection of vintage scientific instruments and apparatuses in her school. Since she liked to place her students' learning in a historical context, she regularly used these objects to illustrate her teaching. Martine did not often use museum resources; nevertheless, she recognized the potential benefit of museum trips to give students an interdisciplinary experience outside the school.

The four participating teachers, two men and two women, had varied profiles. Two teachers taught Secondary Cycle One (Grades 7 & 8), and two teachers taught Secondary Cycle Two (Grades 9–11). One teacher was new to the profession, two were mid-career, and one had over 20 years of teaching experience. Two teachers worked in highly disadvantaged environments and two in middle-class or privileged environments. Despite these differences, they shared an appreciation of the educational potential of museums and had a desire for new experiences that would improve their professional practice.

10.3.2 Reasons for Participating in the Program

The teachers' reasons for participating in the program were varied but for the most part linked to expected benefits for students and other teachers. Xavier hoped that his involvement would help him meet his goal of getting students in his class and school interested in science careers. He also hoped that sharing his experiences would inspire his colleagues to visit museums. Rémi acknowledged that he was following a request from his administrator who asked, "I need someone in science to participate in the development of an exhibition at the *Musée de la civilisation*. Are you interested?" He was already convinced of the impact of his personal and professional experiences on his teaching practice, and he looked forward to this new experience. He hoped that it would expose him to other ways of thinking that would impact his way of doing things in the classroom. Hélène acknowledged that her decision to participate in the program was mainly motivated by the prospect of obtaining free customized visits for her students. Martine became involved in the program out of personal interest; she was curious about discovering the museum from the inside.

10.3.3 Changes in Social Theory of Learning Dimensions

The social theory of learning explains learning as a change in a dimension and as a collective phenomenon experienced by all members of a specific community of practice. Therefore, if one member of a group experiences a change in a dimension, then the community of practice has also experienced a system-wide change; thus, learning has taken place in the system. In other words, it is not necessary for all participants to have experienced the same change; the system has changed, so learning has taken place. In this section, we address each dimension of the social theory of learning independently, keeping in mind that dimensions are always intimately connected, interrelated, and influencing each other.

10.3.3.1 Meaning

For meaning, the first dimension of the social theory of learning, we analyzed the data for moments where teachers expressed having experienced new meanings or a change in meanings, whether related to their professional activity, the museum environment, or their students. All four teachers reported having developed a deeper understanding of the museum and its internal functions and processes. Through visits, lectures, and presentations, they learned about the museum's ways of doing and thinking about schools. Research on museum practices and the school–museum relationship was presented to the teachers. Rémi noted that participation in the program changed his way of looking at museum exhibitions. He said that when visiting museums in the future he would be aware of "all the work behind it—the long months of the design phase, the various professionals involved, the choices made for design, lighting, object placement, and so on."

The teachers reported that their involvement in the design of the science exhibition provided them with new examples of the application of scientific knowledge in various spheres of human activity, transforming their understanding of science as a school subject and as a potential future work environment for their students. As Rémi put it, "It makes our teaching, and certain things we have taught more concrete. In that regard, it can be very positive."

Martine found that the museum exhibition design process echoed the steps of the design cycle that she was required to teach as part of the International Baccalaureate program. Because she had become more familiar with the design process during her immersion in the museum, she now felt more competent in her teaching:

> It's as if going through the process of designing this exhibition helped me take ownership of the whole design process, the whole [International Education Program] design cycle, and it gave me more examples of each step when teaching the program, and more examples to use in a context other than science.

By the end of the program, some teachers were using language specific to museum practice, showing a wider understanding of the museum environment. For example, when Rémi described the behaviour of students in the exhibition halls, he said, "Students are more like butterflies than ants." Butterflies and ants are a direct reference to a well-known ethnographic study (Schiele, 2018) describing the behaviour of indoor museum visitors.

Analysis showed that the teachers' conceptions changed because of their boundary encounter during the program. Whereas before participating in the program only one teacher recognized the cultural value of school museum visits, the end-of-the-program interviews suggested that all four teachers were convinced of the educational potential of museums beyond scholastic usefulness. Furthermore, for the teachers working in remote or disadvantaged areas, school visits were now seen as essential for providing students with a unique opportunity to discover museums and cultural practices they normally would not have had, mostly due to geographical distance. As Rémi said,

> This is the first time they traveled to a museum, so they are 14, 15, 16 years old and have never been to a museum, [and maybe] it ignites a spark in them that says, 'Hey, I feel like going to the museum again to see other exhibitions.'

Martine discovered that going to the museum with her students could have educational benefits other than learning content and could help them become inspired by different or atypical career choices thus expanding their professional horizons. She reflected that museums could interest young people in science careers in a way that stands out from what is already available. In Québec, many organizations offer conferences, workshops, and resources intended to interest students in science. Museums can contribute to these efforts by presenting biographies that make scientists more relatable or by creating interactive exhibitions to arouse curiosity and foster students' motivation toward science. However, Hélène reminded us that not all her students were destined for science careers, so she considered it important to offer a range of learning activities that can interest future scientists and non-scientists alike.

10.3.3.2 Practice

For practice, the second dimension of the social theory of learning, we looked at the changes that occurred in terms of professional practice and the use of tools within that practice. We focused on the teaching practices, strategies, and professional collaborations discussed by the teachers and how they changed over time. They felt that the program had an impact on their classroom teaching practices. For example, Martine recalled using a few themes from the museum exhibition to clarify certain concepts or theories that were more difficult for her students to understand.

Participation in the program convinced the teachers to use school museum visits in teaching sequences that included in-class preparatory and concluding activities. As Hélène noted,

> It will help them, I think, only if there is a preparation phase and a feedback phase. If there is one before and one after the visit, I think it will multiply the impact on young people. Because just the one visit is too short.

For her, the pre-visit activity played a role similar to a film trailer: "It makes you want to go. It makes you want to see the part you didn't see in the preview. It makes you want to explore it."

These teachers emerged from the experience convinced of the importance of visiting a museum with students at least once during the school year. Moreover, they reported having adopted this teaching practice in the year following the program. For example, Hélène made two school museum trips in the year following the program; the year before the program, she had made none. She predicted that museum visits would become part of her teaching. From the outset, Remi considered himself a fan of museums and culture. He reported that participation in the program reinforced this aspect of himself, and he was now convinced of the importance of the museum experience for students, not only as a resource for teaching but also as a broader cultural experience: "I think science culture (I was about to say culture in general) is important, and I want to give as much as possible to my students, because we're in a kind of underprivileged environment."

When these teachers compared their daily reality with their museum experiences, they discovered that things were done differently. At school, everything was done according to a daily schedule and on a short-term basis. Teachers often felt overwhelmed by the hectic pace that prevented them from taking the time to think about and plan their teaching. As Martine articulated, "Everything for us is in the short term, how we try to make concepts accessible, whereas in a museum, the time scale is completely different. It's humbling to see how a concept can be developed over such a long period." As well, any collaborative planning was most often brief due to constraints such as programming and schedules. Working within these constraints left teachers with little space for creativity. At the museum, they were surprised that the research and design for a single thematic exhibition could span several months and even several years. As a result, Martine said that her participation in the program made her want to adopt a similar approach at school to provide students with more creative and satisfying learning opportunities.

10.3.3.3 Community

For community, the third dimension of the social theory of learning, we were specifically looking for new relationships or changes in relationships. The interviews were a little less revealing with regard to this dimension, but we highlight a few examples. Hélène reported that the program provided close collaboration with teachers from other schools and other school boards. She felt enriched by her contact with other teachers and by sharing experiences with the team, whose realities might be different from her own. However, she regretted that the program had not given her the opportunity to produce tools and resources that she could share with other teachers.

Rémi felt that his expertise and knowledge were recognized and appreciated; consequently, he experienced a strong sense of belonging to the teaching community: "being part of a team, creating an exhibition, I've never had that opportunity." As for Xavier, he quickly realized the importance of his role as a representative of teachers to the museum, and this position had the effect of strengthening his feeling of belonging to the community of practice of teachers. Martine said that the beginning of the program was the most difficult for her because she could not understand her role. The teachers initially thought they would play a more important role in the exhibition design process. However, Martine quickly realized that their limited presence at the museum kept them on the periphery and, because her teaching duties occupied most of her time, the research and advisory roles assigned to the teachers suited her quite well.

10.3.3.4 Identity

For identity, the fourth dimension of the social theory of learning, we examined the participants' values, beliefs, and personal stories. Rémi felt that the professional development that he received at the museum over a period of several months required a much higher level of commitment compared to the professional development normally provided by his school board. Being transplanted into a new world was disorienting for him and brought about a new awareness of himself as a teacher. He felt he could become a better teacher if he changed his perspective and experienced the world through the eyes of his students:

> In our professional lives, for me, now, after 23 years … I'm not always faced with situations where I feel I'm lacking knowledge like I felt with the exhibition. I feel I understand young people better and the situations they are often plunged into.

He also appreciated being able to have a long-term relationship with other teachers who shared this experience with him. Rémi quickly identified with the community of practice of teachers and understood the importance of sharing his experience with the other teachers at his school so they could benefit from his experience. He found that participating in the program increased his sense of competency while the museum staff benefited from his knowledge of teenage students. He realized he was able to act as an advisor to the exhibition designers, and this made him proud of his teaching expertise.

In the final interview, Xavier recalled having experienced some uncertainty as to his contributions within the program. He reported that it was a shock for the entire team of teachers with science backgrounds who normally see themselves as analytical and prefer things to be clear and well defined. By participating in the exhibition design process at the museum, Xavier realized that he was not highly creative in his teaching practice: "For large-scale projects like that, we don't really have the time to carry out such projects, with what we have to cover in our subjects. It's less creative here; it's more practical at the school than at the museum."

Hélène declared from the first interview that she felt great personal satisfaction participating in the program. She saw herself as a representative of teachers and an expert on secondary school students: "Of course, we have expertise as teachers, and I basically realized that I have no museum expertise, but my expertise is with young people." Initially, she felt uncomfortable with the unconventional training the program provided. During the first few weeks of immersion at the museum, she had been asked to observe and absorb the museum's culture, but Hélène herself had nothing to do or produce:

> Because, for me, in terms of expectations, at the first meetings, it was really to explain the process, explain the program, and visit the museum's collection, and we did a lot of interesting things. But at some point, you ask yourself: 'Shouldn't you be contributing something?' It was a little unsettling.

However, once she had overcome her initial fears, she was thrilled with what she had discovered. Being behind the scenes of the museum and meeting the many people who worked there fascinated her. Contrary to her school environment, where hours are counted and where projects must be completed quickly, Hélène appreciated being able to step back and observe and reflect. She also felt that the time spent with teachers from other schools and other school boards was helpful; it introduced her to normally inaccessible realities and provided her with professional enrichment.

10.3.4 Discussion

Here we discuss more deeply the changes that the participating science teachers underwent in terms of their stated practices vis-à-vis science education and museums according to the social theory of learning (Lave & Wenger, 1991; Wenger-Trayner et al., 2023). We make connections with the science education literature.

The meaning dimension, which involved understanding of the school–museum relationship, includes the role of aesthetic experience in learning science and how aesthetics relates to the activities in which they are involved (Wickman, 2005). Before participating in the program, the teachers defined the museum as an educational tool to support their teaching. Our analysis showed that this instrumental conception held by the teachers about museums changed as a result of being immersed in the museum's activities. The teachers reported having developed a deeper understanding of the museum and its cultural and aesthetic functions. They now considered museum visits as fundamental to their students' cultural education and not strictly for their scholastic usefulness. Furthermore, another educational benefit of the school museum visit expressed by the teachers was that the students were able to explore professional career choices related to museums.

The practice dimension captured the shift that occurred in the teachers' professional practices. The changes we report are similar to the professional growth noted as arising from contradictions emerging in primary/elementary teachers' activity system during two cycles of collaborative action research (Goodnough, 2016). Our teachers felt that the program had an impact on their classroom teaching practices. Notably, they reported using the themes of the exhibition or their new understanding of the exhibition content in their classroom teaching. Participation in the program convinced them to include school museum visits in their teaching, particularly in sequences that began with preparatory activities and ended with in-class extension activities. In addition, their participation made them aware that it was possible to do things differently and at a different work pace.

The community dimension encompassed the collaborations that developed within the boundary brokers' community of practice, which changed over time. In science education, collaboration is an important context for learning. For example, Meunier et al. (2018) showed that collaborative tinkering can disrupt the practices of scientific education. In our study, extended collaboration with teachers from other schools and school boards was mentioned as rewarding and enriching. The feeling of belonging to the community of practice varied among the teachers; the impact of the program depended on each person's previous experiences.

The identity dimension included the teachers' discourse about themselves, their values, roles and responsibilities, beliefs, and personal stories. Museum practicum experiences can provide teachers with opportunities to develop their skills and disposition for teaching science (Gupta & Adams, 2012). Immersion in a new, unfamiliar professional reality raised a host of feelings; for some of our teachers, participation came with periods of doubt while others felt great personal satisfaction. They felt empowered to act as ambassadors and to represent the interests of teachers and secondary school students.

10.4 Challenges and Recommendations

Here we discuss the conditions for implementing this type of immersive museum education for teachers based on the interview elements that were concerned with difficulties encountered. We report some of the recommendations made by the participants themselves regarding favourable conditions for undertaking this type of program.

A challenge in carrying out the program was that the teachers were given only a few days each year to participate onsite at the museum. This constraint was due to a limited budget and to the difficulty of locating substitutes and organizing travel for the teachers. Despite their willingness to work individually outside the museum, it was difficult for the teachers to keep up with the pace of the program. Each time they returned to the museum, the MEC staff had to update them on the program's progress and the decisions that had been made in their absence.

Another challenge was that the MEC staff had decided to allow the teachers to choose for themselves how to be involved and organize their work together. While Martine felt comfortable with this much freedom, others like Rémi would have preferred more supervision:

> I don't know if it was the same for the others, but it still took a while for us to really figure out where we wanted to go. That was, I'd say, the only hitch about the exhibition and our participation. At some point, we didn't really know what was expected of us.

To address these challenges, Rémi suggested that additional release time be given to teachers so they can contribute more fully to the process and feel part of the museum design team. All teachers recommended that greater supervision and guidance from MEC would benefit future iterations; therefore, it would be beneficial for MEC to provide better supervision, clearer guidelines, and clarification of roles and objectives for the teachers within the exhibition design team to optimize the contribution of outside collaborators.

Our research is in line with other studies exploring aspects of science education including professional learning (Goodnough, 2016), museum university partnerships (Gupta & Adams, 2012), and collaborative action research (Fazio & Melville, 2008). To expand on our results, we propose several avenues for future research. The first avenue might explore preservice teachers' knowledge of museums and their resources from educational and cultural points of view. Results could inform improvements in preservice education, leading to increased knowledge of museum–school interactions, which might allow students to benefit from these cultural resources. A second avenue of research could include offering teacher professional development programs that include immersion in other science museums' communities of practice and documenting these initiatives so that they may be better understood. Immersive museum experiences for science teachers can be relevant for preservice and inservice teacher education at both the primary and secondary level and can lead to learning across all four dimensions of Wenger's (1998) social theory of learning.

Acknowledgements This research was supported by the Fonds de recherche Société et culture (FQRSC) (The social representations of future teachers regarding museums, No: 142159). The authors would also like to thank Pauline Beaudin, project manager and head of the *Centre muséopédagogique* [Museum Education Centre; MEC] of the *Musée de la civilisation* [Museum of Civilization] in Québec City, as well as Dany Brown, director of collections at the Musée de la civilisation, for enabling us to conduct this study. We also extend our thanks to all the school administrators and teachers from the Centre de services scolaire de la Capitale for agreeing to participate in this research project. Finally, we acknowledge the financial and logistic support provided by the Groupe de recherche sur l'éducation et les musées (GREM), Faculty of Education Sciences (FSE), University of Quebec at Montreal (UQAM).

References

Bélanger, C., & Meunier, A. (2012). Partenariat université-musée: Une proposition dans la formation initiale des maîtres [University-museum partnership: A proposal in the initial training of teachers]. In A.-M. Émond (Ed.), *Le musée: Entre la recherche et l'enseignement / The museum: Between research and education* (pp. 253–284). Éditions Multimondes.

Bitgood, S. (1989). School field trips: An overview. *Visitor Behavior, 4*(2), 3–6. http://kora.matrix.msu.edu/files/31/173/1F-AD-10E-8-VSA-a0a2f0-a_5730.pdf

Cohen, C. (2003). Visite scolaire au musée: Représentations d'enseignants en formation initiale en France et au Québec [School visit to the museum: Performances by teachers in initial training in France and Quebec]. In Y. Girault (Ed.), *L'accueil des publics scolaires dans les Muséums, Aquariums, Jardins Botaniques, Parcs Zoologiques* (pp. 195–225). L'Harmattan.

Creswell, J. W., & Creswell, J. D. (2022). *Research design: Qualitative, quantitative, and mixed methods approaches* (6th ed.). Sage.

Daoust, G., Bélanger, P., & Bourret, G. (1974). *L'université dans une société éducative: de l'éducation des adultes à l'éducation permanente* [The university in an educational society: From adult education to lifelong learning]. Presses de l'Université de Montréal.

Dewitt, J., & Storksdieck, M. (2008). A short review of school field trips: Key findings from the past and implications for the future. *Visitor Studies, 11*(2), 181–197. https://doi.org/10.1080/10645570802355562

El Islami, R. A. Z., Anantanukulwong, R., & Faikhamta, C. (2022). Trends of teacher professional development strategies: A systematic review. *Shanlax International Journal of Education, 10*(2), 1–8. https://doi.org/10.34293/education.v10i2.4628

Falk, J. H., & Dierking, L. D. (2018). *Learning from museums: Visitor experiences and the making of meaning* (2nd ed.). AltaMira Press.

Falk, J. H., Dierking, L. D., & Adams, M. (2006). Living in a learning society: Museums and free-choice learning. In S. MacDonald (Ed.), *A companion to museum studies* (pp. 323–339). Blackwell.

Fazio, X., & Melville, W. (2008). Science teacher development through collaborative action research. *Teacher Development, 12*(3), 193–209. https://doi.org/10.1080/13664530802259222

Gajardo, A. (2005). *Entre école et musée: les visites scolaires* [Between school and museum: School visits] (No. 108). Université de Genève.

Goodnough, K. (2016). Professional learning of K-6 teachers in science through collaborative action research: An activity theory analysis. *Journal of Science Teacher Education, 27*(7), 747–767. https://doi.org/10.1007/s10972-016-9485-0

Griffin, J., & Symington, D. (1997). Moving from task-oriented to learning-oriented strategies on school excursions to museums. *Science Education, 81*(6), 763–779. https://doi.org/10.1002/(SICI)1098-237X(199711)81:6%3c763::AID-SCE11%3e3.0.CO;2-O

Grossman, P. W., Wineburg, S. W., & Woolworth, S. (2001). Toward a theory of teacher community. *Teachers College Record, 103*(6), 942–1012. https://doi.org/10.1111/0161-4681.00140

Gupta, P., & Adams, J. D. (2012). Museum–university partnerships for preservice science education. In B. J. Fraser, K. G. Tobin, & C. J. McRobbie (Eds.), *Second international handbook of science education* (pp. 1147–1162). Springer. https://doi.org/10.1007/978-1-4020-9041-7_76

Hutchins, R. M. (1969). *The learning society*. Praeger.

Lave, J., & Wenger, E. (1991). *Situated learning: Legitimate peripheral participation*. Cambridge University Press. https://doi.org/10.1017/CBO9780511815355

Le Groupe de recherche sur l'éducation et les musées. (2000). *Prolégomènes à l'élaboration d'un programme de formation des futurs enseignants à l'utilisation des musées à des fins éducatives* [Prolegomena to the development of a training program for future teachers in the use of museums for educational purposes]. Université du Québec à Montréal.

Martinet, M. A., Raymond, D., & Gauthier, C. (2001). *Teacher training: Orientations, professional competencies*. Éducation Québec. https://www.education.gouv.qc.ca/fileadmin/site_web/documents/dpse/formation_ens_a.pdf

Meunier, A. (2018). L'éducation dans les musées: une forme d'éducation non formelle. In D. Jacobi (Ed.), *Culture et éducation non formelle* (pp. 15–32). Presses de l'Université du Québec (PUQ). https://doi.org/10.2307/j.ctv18pgwvn.4

Meunier, A., & Bédard Daneau, F. (2022). Enseigner et apprendre dans un musée: Vers de nouvelles identités professionnelles enseignantes? [Teaching and learning in a museum: Towards new professional teaching identities?]. In A. Araujo-Oliviera & E. Tremblay-Wragg (Eds.), *Des pratiques inspirantes au coeur de la formation à l'enseignement* (pp. 261–282). Presses de l'Université du Québec.

Meunier, A., & Bélanger, C. (2014). Représentations de futurs enseignants sur l'utilisation du musée d'histoire comme ressource didactique dans l'enseignement-apprentissage de l'univers social au primaire, les sciences humaines à l'école primaire québécoise [Representations of future teachers on the use of the history museum as a didactic resource in the teaching-learning of the social universe in primary schools, human sciences in Quebec primary schools]. In M.-C. Larouche & A. Araujo-Oliveira (Eds.), *Les sciences humaines à l'école primaire québécoise: Regards croisés sur un domaine d'intervention et de recherche* (pp. 149–173). Les Presses de l'Université du Québec.

Meunier, A., Belleville, B., & Grant, O. (2018). Penser avec ses mains ou la philosophie du tinkering [Thinking with your hands or the philosophy of *tinkering*]. *La Lettre De l'OCIM, 177*, 29–35. https://doi.org/10.4000/ocim.2557

Meunier, A., Luckerhoff, J., & Bédard Daneau, F. (2021). L'enseignant, public ou non-public des musées? Paroles de professionnels québécois [The teacher, public or non-public of museums? Words of Quebec professionals]. *La Lettre De l'OCIM, 195*, 30–37. https://doi.org/10.4000/ocim.4359

Ministère de l'Éducation. (2020). *Référentiel de compétences professionnelles. Profession enseignante.* https://cdn-contenu.quebec.ca/cdn-contenu/adm/min/education/publications-adm/devenir-enseignant/referentiel_competences_professionnelles_profession_enseignante.pdf

Musée de la civilisation. (2024). *Institutional profile.* https://mcq.org/en/about-us/institutional-profile/

Popova, A., Evans, D. K., Breeding, M. E., & Arancibia, V. (2022). Teacher professional development around the world: The gap between evidence and practice. *The World Bank Research Observer, 37*(1), 107–136.

Savoie-Zajc, L. (2011). La recherche qualitative/interprétative en éducation [Qualitative/interpretive research in education]. In T. Karsenti & L. Savoie-Zajc (Eds.), *La recherche en éducation—Étapes et approches* (3rd ed., pp. 124–147). Les Presses de l'Université de Montréal.

Schiele, B. (2018). Des *fourmis*, des *papillons*, des *poissons*, des *sauterelles* aux prises avec deux ethnologues de l'exposition. Note sur Ethnographie de l'exposition [Two exhibition ethnologists dealing with ants, butterflies, fish, and grasshoppers. Notes on the ethnographic exposition]. *Communication & langages*, 2(196), 55–72. https://doi.org/10.3917/comla1.196.0055

Wenger, E. (1998). *Communities of practice: Learning, meaning, and identity.* Cambridge University Press. https://doi.org/10.1017/CBO9780511803932

Wenger, E. (2018). A social theory of learning. In K. Illeris (Ed.), *Contemporary theories of learning ... in their own words* (2nd ed., pp. 219–228). Routledge. https://doi.org/10.4324/9781315147277

Wenger-Trayner, E., Wenger-Trayner, B., Reid, P., & Bruderlein, C. (2023). *Communities of practice within and across organizations: A guidebook* (2nd ed.). Social Learning Lab. https://www.wenger-trayner.com/cop-guidebook/

Wickman, P.-O. (2005). *Aesthetic experience in science education: Learning and meaning-making as situated talk and action.* Routledge. https://doi.org/10.4324/9781410615756

Yamagata-Lynch, L. C. (2001, November 8–12). *Community of practice: What is it, and how can we use this metaphor for teacher professional development?* [Paper presentation]. National Convention of the Association for Educational Communications and Technology, Atlanta, GA. (ED470198). ERIC. https://files.eric.ed.gov/fulltext/ED470198.pdf

Zakhartchouk, J.-M. (1999). *L'enseignant, un passeur culturel* [The teacher, a cultural transmitter]. ESF Editeur.

Anik Meunier BA industrial design (University of Montreal), MA museology (University of Montreal), PhD (University of Quebec in Montreal-UQAM) and Doctorate in information and communication science-museology (Avignon University) that earned her the prize for the best Quebec thesis carried out jointly by the Ministry of International Relations and the General Consulate of France at Quebec. She heads the research group on education and museums (GREM) at UQAM and is full professor at the department of didactics in the Faculty of Education and at the Graduate programs in museology in the Faculty of arts. She is interested in the field of cultural mediations, that is to say in the analysis of the professional practices of actors, the methods they mobilize and their effects on the different categories of publics. In 2021, the Faculty of Education at UQAM awarded her the prize for excellence in research, achievements component for her contribution, to date, to the field of museum education.

Charlène Bélanger is Division Manager—Public and Educational Programs at the Planetarium de Montréal, which is part of Espace pour la vie, Canada's largest natural sciences museum complex. She leads a team of 19 people working on the development and implementation of a rich program of immersive films, exhibitions, events and educational activities tailored to the needs of diverse audiences. From 2018 to 2023, she managed PRISM, the digital innovation laboratory at the Montreal Museum of Fine Arts. In this context, she was able to mobilize a vast network of over 300 collaborators who carried out inspiring digital projects contributing to new knowledge in the field of digital mediation. For the past twenty years, Charlène Bélanger has been studying learning in museums, using qualitative, ethnographic and participatory methods. She has thus contributed to several scientific articles and developed training courses for museum and education professionals.

Chapter 11
Exploring the Value of Place-Conscious Pedagogy in Science Education: The Green Crab Project

Darron Kelly and Sharon Pelech

11.1 Introduction

Our research embraces the sociocultural formation and conceptualization of place, which is always intersubjectively mediated, constructed, and contested. Formal education is socially constructed; conceptions of science education, therefore, are functions of culture in the same way that all education systems are intrinsically culture-based. Similarly, conceptions of place are generated and sustained by culture as all understandings of place are intrinsically filtered through cultural perspectives. Recognizing this epistemic presupposition opens possibilities for place-conscious pedagogies that are not bound or based in singular conceptions of place but alive with the varied interests and experiences of educators. Our work is in keeping with the growing interest in "re-theorizing the notion of 'place' in place-based education" (Coughlin & Kirch, 2010; van Eijck & Roth, 2010, p. 895) and explores the educational value of engaging students in local, cultural, and environmental interactions. Whether the aspects of value we identify through our thematic analysis of place-conscious pedagogy cohere with the educational values of other researchers and educators is not for us to decide. We present them as a straightforward, earnest contribution to further understanding and enhancing science education.

D. Kelly (✉)
Memorial University, St. John's, NL, Canada
e-mail: p45dik@mun.ca

S. Pelech
University of Lethbridge, Lethbridge, AB, Canada
e-mail: sharon.pelech@uleth.ca

C. D. Tippett and T. M. Milford (eds.), *Seeing Science Through the Eyes of Canadian Teachers and Learners*, Contemporary Trends and Issues in Science Education 54,
https://doi.org/10.1007/978-3-031-88785-7_11

11.1.1 Place-Conscious Pedagogy

Place-conscious pedagogy occurs when teachers come to recognize the educational potential of the place where they live and begin to use the local community and landscape as focal points for teaching (Gruenewald, 2003; Gruenewald & Smith, 2010; Harasymchuk, 2017; Kelly, 2005, 2006, 2007). Such initiatives take many forms such as restoring school-ground habitats, recording cultural narratives and histories, conducting environmental audits, contributing to population counts of targeted species in the local environment, and learning traditional crafts such as rug hooking and boat building. Insofar as place-conscious pedagogy is motivated by teachers and their understanding of education and place, its many forms are an authentic approach to teaching that moderates and mediates the boundary between schools and local social and environmental settings (Ball & Lai, 2006; Greenwood, 2013). It is also a teaching approach that enhances the agency of teachers to change instructional practices and, in turn, exercise their capacity to positively transform educational experiences for students (Emirbayer & Mische, 1998; Goodman & Eren, 2013; Pelech & Pelech, 2013; Webber & Miller, 2016). This enhanced sense of teacher agency can be shown by "teachers' ending their habitual use of long-held instructional practices, teachers' questioning the meaning and value of existing practices, and teachers' using new practices rather than shielding existing practices from external intervention" (Bridwell-Mitchell, 2015, p. 141). In the place-conscious pedagogy literature, especially rich examples of teachers questioning and changing their practice are given by Coker (2017) who described the educational use of Elon University Forest and how "pedagogies utilizing the forest allow large numbers of students to experience authentic scientific investigations in an interdisciplinary context, leading to meaningful and transferrable learning" (p. 71). The direct process of learning in the social community and/or local environment offers students a diversity of experiences and meaningful opportunities to explore, apply, and advance their conceptual knowledge (Bourne, 1998; Harvard Graduate School of Education, 1999a, 1999b; Liberman & Hoody, 1998; Nichols et al., 2016; Smith, 2007) while engaging as active and informed members of their community and surrounding ecosystem (Basu & Barton, 2010; Deringer, 2017; Henry, 2014; Hutchison, 1998; Wildcat et al., 2014).

11.1.2 Place-Conscious Teaching Practices and Policies

Place-conscious teaching practices that enhance science education can support the achievement of basic curriculum expectations and advance the critical awareness necessary for maintaining the quality of life for future generations (Orr, 2004; Sobel, 2005). Moreover, exploring place-conscious pedagogy may shed light on the overall practices of science teachers in Canada and promote transfer of place-conscious approaches to other countries and contexts. We believe successful science education is as much about the educator and their knowledge of teaching as it is about policy documents and standardized outcomes. Teachers who exercise their pedagogical agency by situating learning in local environmental and cultural settings can provide students with a rich, meaningful, and long-lasting understanding of science—without compromising curriculum standards.

It is encouraging that several provinces have initiated and are implementing comprehensive curriculum redesign that recognizes the importance of place in providing knowledge-rich and meaningful education. For example, in Newfoundland and Labrador, curricular expectations emphasise "common competencies ... that cover a range of skills including self-care, cultural interaction and building a sustainable future" (Government of Newfoundland & Labrador, 2017, p. 65). In Alberta, learning is embedded in relationships, culture, family, Elders, Knowledge Keepers, community, land, connections, memory and history. That educational authorities acknowledge the value of culture and place amplifies the relevance of researching place-conscious pedagogy across the curriculum. Achieving curriculum expectations and developing community and environmental knowledge and agency are important student outcomes and these outcomes are well-documented in the literature on place-conscious pedagogy (Kelly & Pelech, 2019; Lescure & Yaman, 2014; Pelech & Kelly, 2020; Zimmerman & Weible, 2017).

11.2 Research Focus and Procedures

In our work, we examine place consciousness in the specific context of science education in Canada. We also explore the practices and beliefs of Canadian science educators who incorporate place-conscious teaching. We consequently ground our work in two fundamental questions:

1. What practices do place-conscious teachers follow with their science students?
2. How do these teachers comprehend the value of place-conscious pedagogy in the context of science education?

Answering these questions, we theorized, may have important implications for enhancing science education and for fostering environmental sustainability—arguably the most crucial issue of our time.

We explore place-conscious pedagogy in formal educational settings and undertake this exploration through a variety of case studies of elementary and secondary school science teaching in small port and prairie communities in the Canadian provinces of Newfoundland and Labrador, British Columbia, and Alberta. Our research employs in-depth, qualitative interviews and photographic documentation of related pedagogical sites, activities, and outcomes. The interviews and photographs provide a wealth of material for illustrating, explaining, and understanding place-conscious teaching practices. Interview data are processed by thematic analysis (Seidman, 2006) in which we seek to show how place-conscious pedagogy can support teachers and students while providing a source of meaning, hope, and resilience. This thematic analysis connects our central interests to place-conscious pedagogy and science education.

Teachers were recruited via districtwide email messages and were identified through internet searches and in conversations with university faculty, school and district administrators, and teacher association staff. Participating teachers were invited to join the project based on their self-declared use of place-conscious pedagogy in school settings. To date, we have conducted more than forty interviews with teachers in case studies from across three provinces, documenting their place-conscious teaching practices. Our interviews are semi-structured and characterized by open-ended questions that allow for an in-depth, conversational interaction between interviewer and participant (Babbie & Benaquisto, 2010; Klenke, 2008; Seidman, 2006). This type of qualitative interview is response-focused and provides opportunities for participants to explain events and processes as well as express their sense of an idea or concept in relation to their experience (Bryman et al., 2009). In addition to describing their place-conscious teaching practices, teachers are asked to share their understanding of general topics such as student agency, professional identity, assessment, community partnerships, and resiliency considering their approach to teaching. Drawing on their actual practices, teachers choose and frame examples of place-conscious pedagogy and its perceived impact on students, other teachers, and school communities thereby relaying meaningful information in their responses.

In our interviews, the teachers are treated as expert practitioners and encouraged to provide background information about the cultural context of their work, and they are given time to express their ideas in detail (Lofland et al., 2006). This approach creates a positive, collaborative attitude for generating fresh and insightful responses that link place-conscious pedagogy to local community and environmental projects. Interviews are audio recorded and transcribed for analysis. The text is then open-coded; and themes are identified according to the consistency of terms used, topics discussed, and overall meanings conveyed in the interview (Agar, 1980; Creswell & Miller, 2000; Nowell et al., 2017). Repeated themes are independently categorized by the co-researchers. Similar themes are then jointly reviewed, compared, and refined to generate a high degree of confidence and trust in the thematic analysis of responses. In this way, close attention and consideration are given to the possible generation, interpretation, characterization, and development of each research theme (Babbie & Benaquisto, 2010; Bryman et al., 2009; Lincoln & Guba, 1985; Seidman, 2006).

Accompanying the interviews, we use digital photography to document activities, settings, and outcomes and to provide a rich sense of the variety and individual character of place-conscious methods of teaching (Creswell & Miller, 2000; Lincoln & Guba, 1985; Lofland et al., 2006). Together, the interview data and photographic documentation illuminate cases (Agar, 1980; Merriam & Tisdell, 2016; Schensul et al., 1999) that connect place-conscious pedagogy to science education and offer practical illustrations of the value of this pairing. Each case is unique in context and practice but linked by the teachers' place-consciousness and their understanding of the value of places for enhancing science education.

11.2.1 Data Analysis

The starting point for our analysis of the teacher interviews (Nowell et al., 2017) was with the five themes developed from our prior research cases, which showed that place-conscious pedagogy can enrich science education by:

- affirming the professional identity of teachers and personal identity of students
- enhancing the transformative agency of students and the pedagogical agency of teachers
- supporting diverse methods and understandings of teaching, learning, and assessing
- strengthening teacher–student and school–community relationships
- providing a source of meaning, hope, and resilience

Each of these themes is representative of the enhanced educational value teachers who practice place-conscious pedagogy can bring to science education (Kelly & Pelech, 2020; Pelech & Kelly, 2020). As we continue to identify and develop key themes, we are struck by the unique and diverse ways in which place-conscious pedagogy occurs in Canada. This diversity dovetails with Piersol's (2013) observation that there is no "magical prescription" (p. 64) for authentic teaching that can characterize educative practice across all settings. Just as it is culturally informed, place-conscious pedagogy is inherently contextual and infused with local content. Local settings guide teacher agency and shape school–community partnerships. Teachers "listen and dig into the stories of place" as they re-evaluate the meaning of curriculum and pedagogy (Piersol, 2013, p. 64). In this way, place not only acts as a setting for curricular expectations but also offers teachers an opportunity to consider aspects of educational value that extend beyond the requirements of provincial curriculum policy and outcomes. Agency, identity, diversity, relationships, meaning, hope, and resilience have all been expressed by teachers who share their stories of place-conscious pedagogy.

In this chapter, we focus on the place-conscious work of a Grade 10 science teacher on the island of Newfoundland. This study documents a teacher-led initiative that involved students in tracking an invasive crab species and the related activities that the teacher provided. Our main interest is to provide an analysis of how the science teacher comprehended and explained the role of place-conscious pedagogy in the context of science education. We were especially interested in exploring the value of place-conscious pedagogy as a practical means of enriching science education—as a core aim of curricular expectations and as a support for fostering environmental sustainability. After a description of the context and explanation of the teaching initiative, we attempt to show the educational value of place-conscious pedagogy for science education through a thematic analysis of professional identity, pedagogical agency, strengthening relationships, and providing meaning and hope.

11.2.2 Green Crabs on the Burin Peninsula

The *Green Crab Project* took place on Newfoundland's Burin Peninsula, which stretches southwest some 200 km into the Atlantic Ocean. The peninsula is known to be part of the ancestral territory of the Beothuk, the original people of the region. Approximately two-thirds of the way along the eastern coastline of the peninsula is Marystown, which was first established by French and Basques fishermen in the sixteenth century. By the mid-1800s, it was settled predominantly by descendants of English and Irish settlers who continued the local traditions of fishing and shipbuilding. Today, Marystown is a regional hub with close to 5,500 full-time residents, a healthcare centre, shipyard, museum, and high school with approximately 350 students in Grades 10–12.

Our participating teacher was Greg, who was born, raised, and attended school in Marystown and then returned to his hometown to teach secondary school science after completing degrees in natural science and education at Memorial University in St. John's, Newfoundland. At the time of the interview, Greg had recently retired. The Green Crab Project was initiated by Greg after his children found several of what appeared to be European green crabs on Lantern Neck Beach. Greg contacted the local office of Fisheries and Oceans Canada (DFO) whose representatives confirmed that this sighting was the first evidence of green crabs in the Marystown area. DFO (2023) describes European green crab (*Carcinus maenas*) as aggressive and lists it as one of the 10 most unwanted invasive species in the world; it is "highly resilient and can upset the balance of estuarine and marine ecosystems, [negatively] impacting biodiversity … [, can] live 4 to 7 years, eat a wide variety of plants and animals, [and] females can release up to 185,000 eggs once or twice per year" (Impacts section). Thought to have arrived on the Atlantic seaboard of North America around 1817 in the holds of wooden ships, green crabs have steadily spread northward through Canada's Maritime provinces and were first identified in the waters of Newfoundland in 2007. What Greg's children found that day on the beach, however, marked the first time the species had been seen on the southern portion of the Burin Peninsula. Given the threat posed by green crabs to the viability and sustainability of the local fishery, this sighting constituted an important discovery.

Knowing that green crabs were now in the local environment and that the Grade 10 curriculum included a unit on ecosystems and the effects of invasive species, Greg saw an opportunity to engage his students in meaningful learning. He again contacted DFO and asked it there might be a way for his students to collaborate with DFO in gathering information on local green crab populations. "Getting students out [of the classroom] was something we [the science teachers] always looked for," recalled Greg. "They [DFO] said 'yes' and [together we] set up a day where we took all our Grade 10 s—over 100 students —to the beach near my house."

On a cold October morning, when students arrived at Lantern Neck Beach (see Fig. 11.1), DFO scientists, faculty from the Memorial University Department of Biology, and members of the Canadian Auto Workers union—representing local shipyard workers—were there to greet them as they filed off the school bus onto

Fig. 11.1 The inlet at Lantern Neck Beach

the beach. After a safety briefing and instruction on using the nets (see Fig. 11.2), "the students put on hip waders and went out into the water with nets and swept a section of shoreline. They brought everything that was alive ashore," remembered Greg. With supervision and guidance from the adult experts, the students then set about classifying and counting the various creatures caught in the nets, including a fine sample of green crabs. Greg explained, "They [the students] were shown how to measure them [green crabs] and do their age profiles. This information helped put a population number on what was in this particular area. Now DFO had data they could use." For Greg, the most valuable pedagogical feature of that morning on the beach was its experiential nature in that it provided authentic engagement in scientific fieldwork: "[The students] didn't just watch people do science. They took part—they contributed to the science."

11.3 Identification and Discussion of Themes

Our analysis revealed evidence of several of our project-wide themes: affirming professional identify, enhancing pedagogical agency, strengthening relationships, and providing meaning and hope.

11.3.1 Affirming Professional Identity

Greg's understanding of the value of engaging students in local fieldwork developed while he was at university: "I did a degree in geology and every year there was a local field school. I could see that learning was taking place [in the field] that couldn't happen in the building." Having these experiences as an undergraduate student meant

Fig. 11.2 Readying the nets

Greg began his teaching career feeling "very comfortable taking students outside to learn." The Green Crab Project "was an extension" of that sense of comfort and his conviction in the value of experiential learning beyond classroom walls. Fieldwork was an integral aspect of his scientific background and engaging students in the practice affirmed his professional identity as a science teacher. Greg was also well aware of the relevance of experiential initiatives for achieving curricular outcomes, "Fieldwork is mentioned throughout [the Science 10 curriculum guide] so there are linkages there. What we're talking about here is tied directly to the science curriculum." Though his initiative was not motivated by specific sources or studies but drawing instead on first-hand knowledge and experience as a scientist and educator, Greg's observations on the direct and positive link between attaining curriculum expectations and applications of place-conscious pedagogy are well supported in the research literature (Basu & Barton, 2010; Nichols et al., 2016).

11.3.2 Enhancing Pedagogical Agency

Exercising this degree of pedagogical agency is not without challenges. Given the rationalized nature of Canada's school system, any teacher who has tried to substantially diversify their teaching methods, by moving lessons outdoors to engage in local fieldwork for example, will inevitably be confronted by systemic impediments. Our research leads us to recognize and acknowledge the real constraints encountered by classroom teachers while documenting and publicizing the potential for pedagogical agency that exists in place-conscious initiatives (Kelly & Pelech, 2019). With the self-assurance and confidence found at the end of a successful teaching career grounded in a scientific discipline, Greg exercised the agency available to him as a teacher within the provincial school system. Nevertheless, during our interview, Greg was quite willing to talk about the main hurdles to engaging students with the Green Crab Project. First, however, he was quick to note that the school administration had been very supportive of his initiative. From the very start, when Greg asked administrators about partnering with DFO to investigate local green crab populations, the school's administrative team was immediately on board.

Other aspects of the project were less easily navigated, however, and show common system constraints placed on the pedagogical agency of place-conscious teachers. "Scheduling is one of the challenges," Greg recalls, "and supervision [at school and in the field] and substitute time—those are real challenges [to local fieldwork]." Greg found that, with the cooperation of other teachers and administrative support for hiring a substitute teacher for part of the day, he was able to successfully address issues of scheduling and supervision. He also noted that these issues were more readily tackled in the daily operation of a relatively small high school: "Small schools have more flexibility [for rescheduling classes and arranging student supervision]." In this sense, Greg's experience teaching at a small rural high school enhanced his pedagogical agency by giving him opportunity to try diverse methods of instruction like local fieldwork. Greg was able to take advantage of flexibility in the school's structure to support the Green Crab Project.

A final challenge to the project was the administrative requirements of school-district policy—mostly in the form of managing permission slips that had to be sent home, completed, and returned for each student to be allowed off the schoolgrounds. Though a standard practice across Canadian school districts, tracking and collecting over 100 permission slips did add a time-consuming task to the project. This task and other managerial responsibilities fell to Greg as the project coordinator. In considering the extra demands on time and resources, he remained highly optimistic and assured of the value of such place-conscious approaches to science education: "You might get to do it once a year and even then it's challenging ... but it's really, really valuable learning!" During our conversation, Greg suggested that place-conscious initiatives might benefit from having a designated facilitator at or available to the school who could identify, organize, and manage projects in accord with district policy and as part of their employment. He also commented on the potential benefits of using information and communication technology to track and document routine aspects such as attaining parental consent.

11.3.3 Strengthening Relationships

The expense of busing students to the beach was an initial concern, but partnering with DFO, Memorial University, and the Canadian Auto Workers meant the cost of busing and the provision of equipment such as hip waders and nets were covered by the school's community partners. Creating a partnership with local organizations also meant many adult volunteers who understood the value of quality science and could model expectations for safety and scientific enquiry were available to supervise and assist students with conducting the fieldwork. Greg's lifelong ties to the community meant he was able to identify school–community partners that were a good "fit" for the project. Each partner brought an understanding of the need for scientific knowledge and offered various perspectives for its practical application. The partners represented a range of future, science-related options for the students, including postsecondary education and employment.

11.3.4 Providing Meaning and Hope

While Greg expressed a strong sense of professional affinity and conviction for engaging students in fieldwork, he recognizes that such projects may seem intimidating for some teachers: "I wonder how many teachers have the comfort level [to take students into the field]—unless you get out there." He is certain, however, that when given the necessary support, science teachers who "get out there" will find meaning and hope in the educative experiences available through place-conscious pedagogy. When asked about his commitment to situating learning in the local environment and the overall worth of the Green Crab Project, Greg was keen to answer in terms of the project's lasting contribution to positive student experience in science education: "If you talk to any of these students [who have since graduated high school], they would remember that day."

Greg was certain of the deep and lasting effect of the project on his Grade 10 students. "The students did talk about it all the time [in the context of course work]," he observed. The fieldwork on the shoreline and the topic of green crabs as an invasive species appeared in "science-fair projects [and] in the years that followed that experience, it would come up in their Biology class." Perhaps still more indicative of the educational value and inherent meaning of the Green Crab Project was the unexpected depth of the impression it made on students. Even Greg was surprised by the sense of meaning and importance the project held for those students who gathered on the beach that day and the degree to which they identified with the experience. Near the end of the interview, Greg reflected on the Green Crab Project with the hopefulness and perspective of a life spent teaching science:

> High school students are trying to find their way. They are trying to figure out who they are. When you are in the middle of a day like that—out on the beach—and they are doing what they are expected to do … and the next day or a week later you're on supervision duty and there's a group gathered in the hall or sitting on a bench and you hear them talk about it … about what they saw that day or that it was pretty cool. Those are the ones [the lasting impressions] that stand out to me.

From Greg's viewpoint, teaching science by putting students in direct contact with place is powerful and deeply rewarding. It creates a lasting and meaningful connection to learning and the environment for students and a sense of hopefulness for teachers (see Fig. 11.3).

Fig. 11.3 Remember that day... a lasting and meaningful connection with place

11.4 Conclusion

Despite systemic constraints and the need for flexibility and a relaxing of control over the learning environment, Greg insisted there is simply no surer way to nurture a positive disposition for science than to engage students with fieldwork. Though the challenge for teachers is real, a commitment to local fieldwork builds on the spontaneous curiosity of youth. For Greg, initiatives that raise the local consciousness of students and connect them with their surrounding environment are absolutely essential to meaningful science education. "Every student who actually went out there and experienced the cold and the wind and the smells, and felt these creatures ... that had to be something way more than you get in the classroom."

An essential value of place-conscious pedagogy in science education is its source of meaning brought about by connecting and transforming the subject of science to a hands-on way of experiencing and understanding the world and sparking a sustained excitement for discovery. Providing students with opportunities to apply knowledge and develop skills in partnership with local organizations enriches their educational experience as they translate curriculum into fieldwork. Throughout our conversation, Greg expressed a profound sense of the educational value of initiatives like the Green Crab Project for providing students with meaningful impressions of a life enriched by scientific curiosity and understanding. While the discussion provided here explores themes of professional identity and pedagogical agency and the importance of school–community partnerships, we believe it is the inherent link to meaning and hope found through place-conscious pedagogy that is most enriching for science education:

> It's after the fact and when you talk with them years later ... I've had students say to me from twenty years ago, twenty-five years ago, 'Remember that day we did fieldwork and the rain was going sideways and we didn't want to get off the bus?' And they remember that day and they might even remember the things they looked at. That's powerful—they are right from

there and now they're seeing stuff they didn't know existed. They're opening their eyes to something new and was hidden.

References

Agar, M. H. (1980). *The professional stranger: An informal introduction to ethnography.* Academic Press.

Babbie, E., & Benaquisto, L. (2010). *Fundamentals of social research* (2nd Canadian ed.). Nelson.

Ball, E. L., & Lai, A. (2006). Place-based pedagogy for the arts and humanities. *Pedagogy: Critical Approaches to Teaching Literature, Language, Composition and Culture, 6*(2), 261–287. https://doi.org/10.1215/15314200-2005-004

Basu, S. J., & Barton, A. C. (2010). A researcher-student-teacher model for democratic science pedagogy: Connections to community, shared authority, and critical science agency. *Equity & Excellence in Education, 43*(1), 72–87. https://doi.org/10.1080/10665680903489379

Bourne, B. (Ed.). (1998). *Taking inquiry outdoors: Reading, writing, and science beyond the classroom walls.* Stenhouse.

Bridwell-Mitchell, E. N. (2015). Theorizing teacher agency and reform: How institutionalized instructional practices change and persist. *Sociology of Education, 88*(2), 140–159. https://doi.org/10.1177/0038040715575559

Bryman, A., Teevan, J. J., & Bell, E. (2009). *Social research methods* (2nd Canadian ed.). Oxford University Press.

Coker, J. S. (2017). Pedagogy and place in science education. In D. Shannon & J. Galle (Eds.), *Interdisciplinary approaches to pedagogy and place-based education: From abstract to the Quotidian* (pp. 71–83). Springer. https://doi.org/10.1007/978-3-319-50621-0_6

Coughlin, C. A., & Kirch, S. A. (2010). Place-based education: A transformative activist stance. *Cultural Studies of Science Education, 5*(4), 911–921. https://doi.org/10.1007/s11422-010-9290-6

Creswell, J. W., & Miller, D. M. (2000). Determining validity in qualitative inquiry. *Theory into Practice, 39*(3), 124–130. https://doi.org/10.1207/s15430421tip3903_2

Deringer, S. A. (2017). Mindful place-based education: Mapping the literature. *Journal of Experiential Education, 40*(4), 333–348. https://doi.org/10.1177/1053825917716694

Emirbayer, M., & Mische, A. (1998). What is agency? *American Journal of Sociology, 103*(4), 962–1023. https://doi.org/10.1086/231294

Fisheries and Oceans Canada. (2023, January 4). *European green crab.* https://www.dfo-mpo.gc.ca/species-especes/profiles-profils/europeangreencrab-crabevert-eng.html

Goodman, J. F., & Eren, N. S. (2013). Student agency: Success, failure, and lessons learned. *Ethics and Education, 8*(2), 123–139. https://doi.org/10.1080/17449642.2013.843360

Government of Alberta. (2020). *The guiding framework for the design and development of kindergarten to grade 12 provincial curriculum.* https://education.alberta.ca/3575996/curriculum-development-guiding-framework.pdf

Government of Newfoundland and Labrador. (2017). *Now is the time: The next chapter in education in Newfoundland and Labrador.* https://www.gov.nl.ca/education/files/task_force_report.pdf

Greenwood, D. A. (formerly Gruenewald). (2013). A critical theory of place-conscious education. In R. B. Stevenson, M. Brody, J. Dillon, & A. E. J. Wals (Eds.), *International handbook of research on environmental education* (pp. 93–100). Routledge. https://doi.org/10.4324/9780203813331

Gruenewald, D. A. (2003). Foundations of place: A multidisciplinary framework for place-conscious education. *American Educational Research Journal, 40*(3), 619–654. https://doi.org/10.3102/00028312040003619

Gruenewald, D. A., & Smith, G. A. (Eds.). (2010). *Place-based education in the global age: Local diversity.* Routledge.
Harasymchuk, B. (2017). Place-based education & critical pedagogies of place: Teachers challenging the neo-colonizing processes of the New Zealand and Canadian schooling systems. *Environmental Education Research, 23*(9), 1356. https://doi.org/10.1080/13504622.2017.1289501
Harvard Graduate School of Education. (1999a). *Living and learning in rural schools and communities: A report to the Annenburg Rural Challenge.* Harvard University.
Harvard Graduate School of Education. (1999b). *Living and learning in rural schools and communities: Lessons from the field.* Harvard University.
Henry, E. R. (2014). A search for decolonizing place-based pedagogies: An exploration of unheard histories in Kitsilano Vancouver, BC. *Canadian Journal of Environmental Education, 19*, 18–30. https://cjee.lakeheadu.ca/article/view/1288
Hutchison, D. (1998). *Growing up green: Education for ecological renewal.* Teachers College Press.
Kelly, D. (2005). On reconceiving our school grounds. *Pathways: The Ontario Journal of Outdoor Education, 18*(1), 31–34. https://www.coeo.org/wp-content/uploads/pdfs/Digital_Pathways/Pathways_18_1.pdf
Kelly, D. (2006). An ecological conception of school-ground education. *Interactions: The Ontario Journal of Environmental Education, 19*(1), 28–29.
Kelly, D. (2007). Thoughts on content and method in place-based education. *Interactions: The Ontario Journal of Environmental Education, 19*(4), 9–12.
Kelly, D., & Pelech, S. (2019). A critical conceptualization of place-conscious pedagogy. *European Journal of Curriculum Studies, 5*(1), 732–741.
Kelly, D., & Pelech, S. (2020). Canadian contexts for exploring transformative student agency through place-conscious pedagogy. *Education in the North, 27*(1), 110–124. https://doi.org/10.26203/3pbv-8992
Klenke, K. (2008). *Qualitative research in the study of leadership.* Emerald Group.
Lescure, K., & Yaman, C. (2014). Place-based education: Bringing schools and communities together. In C. White (Ed.), *Community education for social justice* (pp. 23–31). Sense.
Liberman, J., & Hoody, L. (1998). *Closing the achievement gap: Using the environment as an integrating context for learning.* State Education and Environmental Roundtable.
Lincoln, Y. S., & Guba, E. G. (1985). *Naturalistic inquiry.* Sage.
Lofland, J., Snow, D., Anderson, L., & Lofland, L. (2006). *Analysing social settings: A guide to qualitative observation and analysis* (4th ed.). Wadsworth/Thompson.
Merriam, S. B., & Tisdell, E. J. (2016). *Qualitative research: A guide to design and implementation.* Jossey-Bass.
Nichols, J. B., Howson, P. H., Mulrey, B. C., Ackerman, A., & Gately, S. E. (2016). The promise of place: Using place-based education principles to enhance learning. *International Journal of Pedagogy and Curriculum, 23*(2), 27–41. https://doi.org/10.18848/2327-7963/CGP/v23i02/27-41
Nowell, L. S., Norris, J. M., White, D. E., & Moules, N. J. (2017). Thematic analysis: Striving to meet the trustworthiness criteria. *International Journal of Qualitative Methods, 16*(1). https://doi.org/10.1177/1609406917733847
Orr, D. (2004). *Earth in mind: On education, environment, and the human prospect.* Island Press.
Pelech, S., & Kelly, D. (2020). Teacher identity and agency: Learning and becoming through place-conscious pedagogy. In E. Lyle (Ed.), *Identity landscapes: Contemplating place and the construction of self* (pp. 210–221). Brill | Sense.
Pelech, S., & Pelech, W. (2013). Shifting paradigms in secondary science classrooms: Teaching and learning from an ecological perspective. *The International Journal of Interdisciplinary Educational Studies, 7*(4), 41–51.
Piersol, L. (2013). Local wonders. In D. B. Zandvliet (Ed.), *The ecology of school: Advances in learning environments research* (pp. 63–72). Sense.

Schensul, S., Schensul, J., & LeCompte, M. D. (1999). *Essential ethnographic methods: Observations, interviews, and questionnaires.* Alta Mira Press.
Seidman, I. (2006). *Interviewing as qualitative research: A guide for researchers in education and the social sciences* (3rd ed.). Teachers College Press.
Smith, G. A. (2007). Place-based education: Breaking through the constraining regularities of public school. *Environmental Education Research, 13*(2), 189–207. https://doi.org/10.1080/13504620701285180
Sobel, D. (2005). *Place-based education: Connecting classrooms & communities.* Orion Society.
van Eijck, M., & Roth, W.-M. (2010). Towards a chronotopic theory of "place" in place-based education. *Cultural Studies of Science Education, 5*(4), 869–898. https://doi.org/10.1007/s11422-010-9278-2
Webber, G., & Miller, D. (2016). Progressive pedagogies and teacher education: A review of the literature. *McGill Journal of Education / Revue Des Sciences De l'éducation De McGill, 51*(3), 1061–1079. https://mje.mcgill.ca/article/view/9315
Wildcat, M., McDonald, M., Irlbacher-Fox, S., & Coulthard, G. (2014). Learning from the land: Indigenous land-based pedagogy and decolonization. *Decolonization: Indigeneity, Education & Society, 3*(3). https://jps.library.utoronto.ca/index.php/des/article/view/22248
Zimmerman, H. T., & Weible, J. L. (2017). Learning in and about rural places: Connections and tensions between students' everyday experiences and environmental quality issues in their community. *Cultural Studies of Science Education, 12*(1), 7–31. https://doi.org/10.1007/s11422-016-9757-1

Darron Kelly, BA (Mount Allison), BEd (Queen's), MEd (Queen's), PhD (Memorial) is an Associate Professor with Memorial University, Faculty of Education. Darron is an award-winning, SSHRC-funded scholar who examines applications of critical social theory in educational administration (including communicative rationality, transformative leadership, and moral policymaking) and explores the educational value of place-conscious pedagogy (including teacher identity, school-community partnerships, and transformative student agency).

Sharon Pelech, BEd (University of Alberta), MEd (University of Lethbridge), PhD (University of Calgary) is an Associate Professor of science education and Associate Dean of Undergraduate Programs in the Faculty of Education at the University of Lethbridge. She teaches at the undergraduate and graduate levels in curriculum studies and science education. Previous to this, Sharon spent over 18 years as a secondary science teacher in the Northwest Territories, Northern Alberta, and Calgary. Her experiences as an educator in a variety of locations has influenced her research interests in science education, ecopedagogy, place-conscious pedagogy and interpretive (hermeneutic) research.

Chapter 12
Building New Relationships Through Informal Science: Case Studies of Urban Science Clubs and an Inuit-Led Water Stewardship Project

Jrène Rahm

12.1 Introduction

In this chapter, I offer insights into possibilities for science education projects grounded in informal science practices as I explore how rich learning opportunities can be when viewed from the perspective of relational and Indigenous epistemologies of science and the world. In doing so, I build on the literature on science education at the intersection of formal and informal education understood from an asset-based orientation and interrogate what it means to learn and become in a space or activity that is experienced in empowering ways by its participants (Nasir et al., 2021). At the same time, I recognize that in the current landscape of science, technology, engineering, and mathematics (STEM) education the rich cultural repertoires and experiences of historically minoritized communities are still too often ignored or silenced, given on-going practices that uplift and are guided solely by Eurocentric visions and norms about doing and becoming in science. It has led many authors to critique current practices of exclusion and to engage deeply with new imaginaries of equity and social justice-driven science practices and what *rightful presence* in STEM learning might imply—the latter term suggesting that all students have a fundamental right to become legitimate members of science practices without having to pretend to be somebody they are not (Tan & Calabrese Barton, 2023). I take these ideas further, conceptually, as I draw on an emergent literature viewing learning and identity as relational and then engage with the epistemological shifting called for by such a

J. Rahm (✉)
Université de Montréal, Montreal, QC, Canada
e-mail: jrene.rahm@umontreal.ca

C. D. Tippett and T. M. Milford (eds.), *Seeing Science Through the Eyes of Canadian Teachers and Learners*, Contemporary Trends and Issues in Science Education 54,
https://doi.org/10.1007/978-3-031-88785-7_12

view. In doing so, I offer ideas on how to reimagine science education at the intersection of formal and informal settings that embraces and implements justice-oriented pedagogy and practice.

In this chapter, I draw on qualitative data from two projects: one involving urban science clubs and one involving a community-led water stewardship program. The first project, Science Youth in the City, involved two cases of science clubs in Montréal high schools in ethnically diverse, underserved, urban communities. The second project, Water Stewardship, was a single case focused on the educational value and richness of community-led youth programs in Inuit Nunangat. My analysis highlights science practices that are locally relevant, transformative and empowering, and deeply committed to social justice. I end with a discussion of the implications of a relational perspective for navigation among epistemologies and science practices, which I contend is at the heart of an equity and social justice-driven lens to informal science education.

12.1.1 Learning and Identity as Relational

A relational view of learning and identity is foundational to sociocultural theory; it is grounded in the seminal work of Vygotsky (Rieber & Carton, 1987) and extensions of it in the social practice theory of Lave and Wenger (1991) and the figuring of worlds and selves by Holland et al. (1998). A relational view is also central to current posthumanist critiques and research approaches in education (Marin, 2020; Taylor, 2016) and the learning sciences (Takeuchi et al., 2023). I understand learning and identity as being entangled with the world and happening in a physical place yet also as part of a larger learning ecosystem (Tuck & McKenzie, 2015). Accordingly, learning and identity can be understood as locally situated and simultaneously entangled within systems of formal and informal education. Such a view centres multiple space–time relationships (e.g., personal histories) alongside current and future meanings and selves in science (Rahm et al., 2022). Emphasis also rests on a decentering of humans through a focus on deep connections with the more-than-human elements of place (Hecht & Nelson, 2021) and with attention to nature-culture relationships (Marin, 2020). A relational view resists the objectification of learners and educators, calling instead for a focus on human interrelationships and creativity in recognition of possibilities and transformations that uplift students and flatten power differentials, which can often undermine learners' full participation and success in science (Rahm et al., 2023). Pedagogical relationships are forged over time, resulting in the deep affinities (Ehret et al., 2018), mutual respect, and reciprocal relationships that Nxumalo (2020) called "a flourishing together" (p. 563). This flourishing implies the circulation of bodies and materials, in and beyond place, and their multiple entanglements in emergent social relationships. Such a perspective builds upon more traditional educational research and practice and focuses on "epistemological, ontological and axiological expansions" (Takeuchi et al., 2023, p. 260). A relational view shifts epistemological assumptions that I address in the three cases in this chapter.

12.1.2 Relational Stance Through a Desettling of Epistemologies

A relational stance calls for the unsettling of ways of knowing and challenges colonial ideologies that mark science education by putting emphasis on the existence of a single true science and the Western way of knowing and becoming in science (Bang et al., 2018). A relational stance situates ways of knowing as emergent from relationships (Wilson & Hughes, 2019); subjective, multiple, and heterogenous; embedded in cultures, communities, and lived experiences (Adams et al., 2020). This stance calls for a "multiplication of knowledges-sciences"(p. 1504), and values multiple ways of knowing emerging from a sense of place through relations.

A relational stance can be found in all three cases. The first two cases arise from a single project and build on an epistemological footing that values multiplicity and dignity and, in light of relationship-building, centres lived emotions and embodiments in science. The third case comes from a second project that speaks to Indigenous science and knowledge systems, which are living and changing systems of knowing (Cajete, 2000; Lipe, 2019). The inherent relational view is somewhat different between the cases, yet this epistemological stance provides a rich complementarity to transformative notions of learning and identity in science (Bang et al., 2018).

As such, I provide support for the argument that "heterogeneous sciences – specifically Indigenous sciences – can expand our collective knowledge and are critical if sciences (in their plurality) are to become champions of the common good and adequately respond to contemporary problems" (Bang et al., 2018, p. 151). Indigenous science is understood as emergent from subjective experiences and respectful, reciprocal relationships with nature and the world, implying a lifelong journey of coming to know and a storying of the world (Cajete, 2000). Indigenous science is driven by the need to contribute to the common good while keeping all relationships in balance.

12.2 Overview of Methods

Taking an ethnographic approach in the two projects, I relied on a combination of video ethnography and ethnographic observations. These data sources were supplemented by semistructured interviews of key participants and small group interviews by the research team (author of this chapter along with graduate students). Data collection was pursued over time in each project and each case, resulting in a vast dataset of video, observational fieldnotes, artefacts, individual and group interviews.

Video data collected for Project 1 was reviewed by the research team and then subjected to interaction analysis (Jordan & Henderson, 1995). The first round of analysis was conducted individually and focused on interactions in the data set that offered insights into moments of meaning making in science and individual-material entanglements; second, by sharing such telling moments in research team meetings, we identified recurrent patterns across the data set; third, those moments of interactions were transcribed and enhanced with screenshots to capture entanglements of bodies with materials. The results of that analysis were further enriched through a thematic analysis of the interviews, which was the primary source of data and analysis pursued for Project 2. The final step implied a bricolage of these multiple data sources and crafting of vignettes that tell stories of key moments of engagement with science, a selection of which are shared in this chapter.

12.2.1 Project 1: Science Youth in the City

The Science Youth in the City Project involved two cases of high school science clubs that operated in Montréal from 2011 to 2014. The clubs were co-designed with a team of graduate students in education and science and community partners including science teachers and pedagogical counselors. Case 1 focuses on a video documentary project at School A, pursued in a winter term, with nine students (seven girls, two boys) aged 13 to 14 years. Seven participants were first generation immigrants who had been in Canada 3 to 10 years, while the other two students were born in Canada to Canadian parents. Two graduate student instructors (one in science, one in education) worked with the students once a week during lunch break with a special arrangement with the science teacher and school principal when we opted to pursue fieldtrips. Case 2 focuses on six students (one girl, five boys) aged 13 to 14 years, all of whom were first generation immigrants who had been in Canada 3 to 10 years. These students were members of School B's science club and took part in a 2-week summer internship at the local Botanical Garden, travelling there daily with a graduate student mentor.

12.2.2 Project 2: Water Stewardship in Nunavut

Case 3 arose from a second project that was part of an initiative committed to the documentation of the educational value and richness of community-led youth programs in Inuit Nunangat. The Water Stewardship Project responded to community concerns about drinking water quality in Mittimatalik (Pond Inlet), a community situated on the northeastern shore of Baffin Island in Nunavut. The project was initiated in 2011 and led by Tim, who was later joined by co-lead Trevor. Tim is Inuk and was born in Iqaluit, Nunavut, and later moved to Pond Inlet, Nunavut, the community where this project took place. Trevor, also Inuk, was born in Pond Inlet. Together with a southern community organization, they were guided in developing water stewardship skills and methods, blending Indigenous knowledge systems with Western science, through joint work with four local youth during the spring and fall of 2015 and 2016. While other data were collected until 2018, in this chapter we focus primarily on the latter years when water data sampling was pursued together with two youths, Mikael and Eric.

12.3 Descriptions of Three Case Studies

12.3.1 Case 1: Creating a Video

At School A, the nine student members of the science club met once a week throughout the school year, with the goal of producing several science documentaries. In the second half of the school year, following extensive brainstorming on biodiversity, further supported by a fieldtrip to the Biodiversity Centre at the Botanical Garden (https://espacepourlavie.ca/jardin-botanique), we eventually settled on the theme *The Impact of Invasive Species on Threatened Species in the Urban Setting* for the second video. The students divided themselves into three teams; each worked on a subtheme—urban ecosystems, threatened species, or invasive species—and developed a script. An ecologist visited the club, and each team prepared questions to ask him for their part of the documentary. The interview was filmed; each team used that footage, in addition to text that they created and images that they had taken during their visit to the Biodiversity Centre or had found online to create the video's content. This content was then shared with the whole group for feedback. Some student–instructor exchanges follow.

Yousra: There is not enough written text, we rely too much on the verbal from the interviews.

Marc: The images go by too quick; I had no time to read the text. [laughter in the back]

David [an instructor]: Alice, it's maybe a bit too long, too much interview data and not enough text. [Fieldnotes, March 2013]

The feedback guided each team's editing process as they refined timing of text and images, which led to extended negotiations among team members as they figured out how their science content was best woven together.

Lydie: Ok, so now we are working on invasive species, I think? Is that it?
Yousra: Mmmh, yes, we will have to be brief.
Lydie: Again? We already got two, three, four, five, six text entries on that.
Inese: You did how many seconds per image?
Lydie: We're doing one minute and thirty.
Inese: Five?
Lydie: No, more than that.
Inese: 39
Lydie: No, more than that.
Inese: But that's what's written here.

Lydie: Yes, but we'll fix that, so we will fix it, that it makes 9 seconds, then it will be more.
Inese: 9 seconds?
Lydie: Yes, because of the text, it's long, so you can read it.
Yousra: The Introduction?
Inese: You kind of wrote too much, maybe.
Lydie: By the human, involuntary.
Yousra: Mmmh ... no!
Lydie: Do we write, voluntary or involuntary, or something like that?
Yousra: Hey, this is great, at least now, I really understand! [Video transcript, March 2013]

The students co-constructed the video, responding to the critiques they received and timing the text with the images, to arrive at the telling of a sensible science story. They contributed to the video production in different ways and began to see themselves as video producers. At the same time, they struggled to understand the complexity of urban ecosystems and their link to biodiversity. Although the fieldtrip to the Biodiversity Centre helped to some degree, students seemed rather detached from nature in their community. That detachment prompted the club leaders to propose a fieldtrip to a nearby park along the St. Lawrence River.

The participating students were very invested in the second fieldtrip; they had digital cameras to capture anything they deemed scientifically pertinent to their video project. They observed either the water for signs of fish or the soil, exploring different kinds of plants and species. Initially, they relied on guidance from the two leaders but then ventured off and made their own discoveries. The fieldtrip was a nice break from the tedious editing work. With help from the leaders and books, they identified birds, tried to understand damage to trees, and eventually found a salamander under a rock. Kassandra took pictures of an empty bird nest (see Fig. 12.1) to share with their science teacher. Students were amazed at the amount of plastic and other polluting materials that were interwoven with leaves and branches, making for a beautiful and solid nest. It made them appreciate the reality of pollution in an urban space close to their home.

Once back in the classroom, one leader (Ivan) helped students make a connection between the plastic and other polluting debris in the park and the documentary on biodegradable plastic that some students had co-created the previous fall:

Ivan: Alice, you did the video documentary on waste, and you made the comment now, that there is a lot of waste and pollution in this park; do you think it has a big impact on the animals in this park?
Alice: Yes. There is a bird who made his nest with dog muck.
Ivan: Oh yes, so you have seen the impact, but we do not know if that affects birds in any way, right?
Kassandra: Well, the bird did an excellent job integrating these wastes into the nest, there was even a string (piece of rope from boats probably from the river), the string was like attaching the nest to the tree. It was really interesting to see. [Fieldnotes, April 2013]

Fig. 12.1 Student photos of a bird nest

To see with their own eyes, to touch and explore an urban environment in the vicinity of their school and home turned out to be an important means to contextualize the video project and make it locally relevant. Many of the pictures were used for an arts project depicting biodiversity in the urban context and exhibited in a library, which included an opening event and screening of their documentary that was attended by their school director, library personnel, and some of their peers. Figure 12.2 shows a timeline of an excerpt of the video made by the School A group to showcase the impact of invasive species on threatened species in urban spaces.

School A: *The Impact of Invasive Species on Threatened Species in Urban Space*
Excerpt below 2:15min (total length 7:25min); all along same electronic music typical of science documentaries; text (at times multiple ones following one another) & scientist offer explanations and science content.

Introduction to Science Club: What is it, why do you participate in it? A slide of text: 'club implied the participation of 9 youth'	Title of movie: *The Impact of Invasive Species on Threatened Species in Urban Space* *1. Urban Ecosystems* Marc & Mellan *2. Threatened Species* Ines, Lydie & Yousra *3. Invasive Species* Lara, Marina, Alice & Kathy Do you know what an urban ecosystem is? It's about species that live in the city	Urbanisation changes the biodiversity of plants via different mechanisms like the transformation of the habitat and changes of the environment through the introduction of species.	Talk of green corridors that help species move about in urban spaces. Threatened species (title): *Threatened species are endangered species.*	Interview of scientist: 'Can you think of some examples of endangered species in the urban context?' Clip from interview with scientist; his answer is then followed by a summary in text form and picture from fieldtrip of hands of youth with text about human impact on the environment.
0:43min	0:24min	0:21min	0:08min	0:98min

Fig. 12.2 Timeline of an excerpt of the 7:25-min video documentary that students produced

12.3.2 *Case 2: Attending Summer Camp*

Science club members attended a summer camp that provided opportunities for building new relationships with nature and others in the urban space of the Botanical Garden. The intention was that students would be camp assistants for the younger children taking part in the summer camp. The following journal notes by Mayra, the university student mentor who accompanied the science club students, offer some insights into the summer camp:

> Today was our first day and it was so interesting to see how the children in summer camp in the Botanical Garden interacted with the youth from our science club. What I noticed most is the contrast between them in terms of their cultural background and socioeconomic status. On the one hand, the campers are dressed and act in ways that gives them away as primarily middle- and upper-class and well-educated Québecers, with little visible ethnic diversity, speaking an impeccable French, nicely dressed, calm and eager to perform well and be noticed. In contrast, our youth from the science club struggle to express themselves well in French, are very shy in the garden when asked to speak up in front of the others but laugh a lot and at times are off-task and noisy when among themselves. They use a lot of gestures and are quite fraternal. [Journal notes, Mayra, July 1, 2013]

After a day in the program, most of School B's science club students took on the role of camp participants rather than assistants, given how new the camp environment was for them. Few had participated in summer camps before. They were eager to learn as much as possible from the many activities they experienced each day. They became keen observers of the environment and had many questions, maybe in part due to their prior video documentary work. Interestingly, it was Mayra who often ended up mediating the relationship building the land called for and not the camp instructors, as the following journal notes suggest:

> The campers were eating blackberries directly off the bushes in the blackberry patch today. Our three youth were sitting with me, somewhat discouraged from the heat in the garden. I encouraged them to go and taste the blackberries. They told me they never tasted them before but would be interested in finding out how they taste. I asked one of the campers, a young girl, to bring us some blackberries to taste. She was all excited to share and watch us taste them. The camp instructor was shocked to learn that our youth had never tasted blackberries: 'What? They don't know what a blackberry is?' At first, Kyle, one youth, refused to eat anything that came from a tree or a bush in the garden. Yassine, another youth, was worried that it might be poisoning. In the end, curiosity took over and he tasted it anyway. Moustapha, another youth, then spent the rest of the day going back, eating more blackberries. He even asked to bring some home for his mother. [Journal notes, Mayra, July 1, 2013]

Our students experienced many things that were new to them besides eating plants. For instance, the eco-arts group created art with whatever objects they could find in the forested area of the garden, a display that was then shared with the other campers (see Fig. 12.3, left). The ecology group went fishing and observed insects, leading to a science fair project at the end of the camp. The forest group learned about ways to survive in the forest and how to find edible plants (see Fig. 12.3, right). They also constructed a shelter in the forest and met with an Elder from the First Nations Garden who shared stories about the relationship between land and food. The

Fig. 12.3 Found art created by the eco-arts group (left); Forest group walking in the forested area (right)

birding group spent time observing and identifying birds and met an ornithologist. The gardening group came to enjoy the connection with the land by planting plants, weeding, touching a frog, catching insects, or learning the names of the birds.

The summer camp experience engaged the students in land-based education that helped them to develop a new sense of self in relation to place, which they summarized as follows during a brainstorming session following their internship:

- I felt free, without electronic gadgets and apart, and away from the big population of the city.
- I felt like I was in nature.
- I felt great with the children and being in nature.
- I felt incorporated and part of nature.
- I enjoyed being with the children from the camp, and I learned a lot about ecology and art.
- I felt included in the group and in the activities; the campers had fun with us. [Fieldnotes, July 2013]

We were somewhat surprised by the students' positive comments because we had sensed a disconnect between them and nature, between them and the younger campers, and between them and the camp instructors that we met through the different activities in and beyond the Botanical Garden, as the following notes by Mayra suggest:

> Today, we left the Botanical Garden to visit an urban eco-centre nearby. These centres offer green resources to its communities, such as composting, recycling, and workshops tied to community action projects. As a group, we walked a maze of alleys, alleys that were transformed into green spaces, a recent project within that particular community in Montréal. The two instructors talked at length about urbanisation and the greening of cities in a terminology borrowed from studies in urbanisation and geography. It was challenging for youth to follow along given the language and, naturally, we started a game of identifying plants we found in the alleys. Later, I found a plant full of berries and had a taste. It got most [students] interested in tasting the berries for themselves. It led to a conflict, however, with the staff of the eco-centre. They did not think of green alleys as places to eat, arguing the berries were there to stay and attract birds. I tried to explain that the goal of our summer

> program was to develop youths' curiosity and interest in biodiversity and Indigenous plants through the senses and, hence, we thrive on tasting sessions in our program. The instructors still asked us to stop and then continued to talk about Indigenous plants in the alleys which they compared to the old houses and cultural heritage of the neighbourhood. [Journal notes, Mayra, July 3, 2013]

This example attests to the irony behind *hands-off* approaches to environmental education as described by Sobel (2012):

> Much of environmental education today has taken on a museum mentality, where nature is a composed exhibit on the other side of the glass. Children can look at it and study it, but they can't do anything with it. The message is: Nature is fragile. Look, but don't touch. Ironically, this 'take only photographs, leave only footprints' mindset crops up in the policies and programs of many organizations trying to preserve the natural world and cultivate children's relationships to it. (p. 2)

The green alleys were presented as exhibits of nature not to be touched or interacted with, while the Botanical Garden and its programs—as well as our science clubs—encouraged exactly the opposite and invited students to observe, touch, eat, and actively interact with, create, and engage with nature through multiple senses. By doing and interacting, students were given the opportunity to build new relationships with nature in their community, learn when embodied ways of engaging with plants were appropriate, and understand science as a relevant part of their lives. Building those kinds of relationships were key to developing their science literacy, which became deeply grounded in and related to their lives.

12.3.3 Case 3: Mittimatalik Water Quality

Case 3 took place in Mittimatalik (Pond Inlet), a community that relies on water delivery as is typical in many rural communities in Inuit Nunangat yet a challenge in terms of water quality (Medeiros et al., 2017). Elders referred to the addition of fluoride and still others recalled the frequent boil-water advisories, which are measures put in place when water quality is compromised. The Water Lake serving the community was created during the 1970s settlement period; the community, however, has increased substantially in its population and water needs; to rely on one water source can be risky. These water management practices stand in stark contrast to the Inuit's traditional practices prior to settlement. Through careful observation of indicators like taste, colour, flow rate, and turbidity, people would monitor water quality and Elders recalled few incidents of water-related sickness.

The Water Stewardship Project emerged within this context. Tim, the project leader, who was later joined by Trevor as co-leader, explained:

> We wanted to develop a project that would provide us with the opportunity to conduct serious research and answer the preoccupations of our community in a way that would build our skills and knowledge for the benefit of our community. We found out that the best way to achieve these goals was by taking the lead in research and manage it ourselves and, when needed, request the help of researchers in universities and NGOs. I started this project given

> encouragement by community members and Elders and wanted to respond to the concerns of community residents who complained of stomach illness. Our community had no equipment and no capacity back then to research water. [Interview, August 2015]

This project contributed to the development of "community power for self-determination" (Johnson & Nelson-Barber, 2019, p. 3). It also relied upon a blending of Inuit knowledge systems with approaches from Western science. Participating youth and the project co-leaders became apprenticed in water monitoring practices through training by scientists. Based on Elders' advice, the team sought out and sampled water lakes further away, assessing their quality and potential as future additional water sources (see Fig. 12.4). There was also a land camp with Elders to learn more about water practices and concerns, knowledge that was documented and shared with incoming scientists to further enrich the local water stewardship practices (Anaviapik-Soucie et al., 2015).

The project offered youth an immersion into culture-based learning, given its promotion of Indigenous knowledge systems, language, and culture through exchanges with Elders and the two project co-leads. Youth learned about locally

Fig. 12.4 Water sampling at lake (top left); Waterflow sampling in river (top right); Students walking and observing (bottom left); Analyzing water samples in the local laboratory (bottom right)

and culturally grounded water practices resulting in "intergenerational out-of-school connections, [as well as intergenerational connections in support of] learning for a real-world purpose" (Johnson & Nelson-Barber, 2019, p. 24). That is, to gather water for Elders on the land or from an iceberg is not simply about water; it is about rebuilding relationships with the land and with each other for the common good. Youth learned how time consuming and challenging it can be to gather and transport ice blocks to the community so as to ensure community wellness and that Elders have access to *tasty water*.

Mikael was thinking of pursuing the Environmental Technology Program offered through Arctic College but was advised to wait a year, given that his high school grades were not that strong. Tim approached him that summer to become one of the research assistants. "I got really excited, just wanted to learn and understand and was hopeful it'll help me through the years to come" [Interview, Mikael, September 2015]. After participating, Mikael reflected:

> I wasn't really much of a hunter kind of guy, so I had the difficulty of not knowing as much what to do when we were going out to the remote places to get water samples. But it was an amazing experience, I've learned a lot, ... about land, about the animals and about our rivers. Seeing the patterns from how it went when it was clean, and when it wasn't so clean. … we walked a lot along the rivers, cleaned up, like, we were checking to see what was in the river or what might have contaminated the river. It was an awesome time, checking to see all the garbage around and taking the garbage out from the river. And we saw a difference in the water quality measures when we took all the garbage out from the river. So, it was pretty cool to know that just garbage that people leave around in the rivers does make the river like not so clean. [Interview, Mikael, September 2016]

The project helped Mikael reconnect with the land and come to know anew the manner everything is interrelated. It made him "put my own garbage in my pocket" and care for the environment in new ways. He valued the many opportunities the program offered, whether it was tracking the E.coli over time, thinking about what influenced the water quality, or exploring the potential of a water lake further away from town that the Elders had recommended. For him, these experiences helped him "see the land in all its glory, like, this is where I grew up, but I've never seen it in this light before, in this way."

Mikael also had an opportunity to reconnect with Elders during the land camp. He recalled struggling to connect with the Elders initially; but once he was asked to help out more in camp, he was able to build relationships with Elders in ways he highly valued:

> I started making them tea, I helped set up their tent for them and I carried a lot of the stuff for them. So, it was, I don't know, it was just an amazing experience just talking to them, and doing stuff for them and then knowing that ... like I told them about myself, so them knowing, getting the idea of who I am, helping me out, like telling me to do things, and me just working and helping them out with literally anything I could do … I felt good. [Interview, Mikael, September 2016]

The Elders respected him and saw him as a youth who tried; this was a humbling new experience for Mikael who felt that in the past he was put down too often, which then discouraged him from trying. He found himself in a vicious cycle that the words of the

Elders broke since they "told me they were proud, and they told me to keep my head up and to ignore anyone that puts me down." Engaging with Elders in dialogue and being with them on the land was healing for Mikael. Learning became reciprocal, an opportunity for Mikael to share his ways of life with the Elders, grounded in respectful relationships and deep listening:

> Telling them what I know, and what they didn't know. I felt that gap in generations where, back in the day, they don't know, now, and what we do know now, and teaching them, making them catch up with what we are capable of doing nowadays, through the study of water, helping them see what we do, to learn more. They were really interested about it, and I guess it felt good for me because I felt useful. Again, I felt useful that I could teach someone with a lifetime of experience to learn something new in their life and something they've kind of grew up with but didn't really take note of. So, learning about it, it felt pretty great, I felt pretty great. [Interview, Mikael, September 2016]

Being on the land with Elders fostered feelings of calmness, balance, peacefulness, and happiness among everybody in the group, contributing to Mikael's feelings of connections and well-being.

Being with Elders was also a means for youth to re-engage with the land and each other in Inuktitut and, in doing so, live in and through Inuit knowledge systems in ways not accessible elsewhere, as Eric suggested:

> I like talking to Elders, listening to their stories and getting answers. And I love talking to Elders most of the time because they know anything. And I like talking Inuktitut a lot. Really good, because I'm Inuk and it's my language. I've learned English at school from white people and it's kind of harder to understand in English. I like going camping, hunting, especially with Elders. They would teach us a lot. They've experience going camping and hunting, ... they would know how to survive. [Interview, Eric, September 2016]

This quote hints at Indigenous spirituality—"a way of understanding the interrelatedness of forces and elements found in the world. It is the foundation of Indigenous science" (Lipe, 2019, p. 474)—something Inuit have always done and still do through environmental stewardship and being on the land (Tagalik et al., 2023).

Tim explained that too often, the participation of Inuit in research is encouraged to get funding or to look good during a presentation where Inuit then serve as "poster boys and butlers." But few opportunities exist where Inuit lead environmental stewardship projects and assume all steps of a project. As Eric noted during participation in a conference, "we need more community-based researchers to come and present their work and with their emotion tell the audience how they feel." Talking about their own land and initiatives as Inuit is empowering but also crucial to break ongoing colonisation in Northern research (Tagalik et al., 2023). Eric presented in Inuktitut at the scientific meeting "because it was easier for me to say the words that I can't really explain in English and other people who listen or understand Inuktitut would understand what I am saying." He hinted at the manner that language and ways of knowing are intertwined. His comments also suggested that by "using Indigenous knowledge systems, learners [like himself], develop repertoires of practice that help them function effectively as they make sense of real-world experiences" (Johnson & Nelson-Barber, 2019, p. 1353). These repertoires not only constitute lifelong learning and future stewardship of the land but also ensure the keeping alive and regeneration of Inuit knowledge systems that are key to community well-being and self-determination.

Case 3 speaks to stewardship of the land, the building of relationships with the land and with Elders, and Inuit knowledge systems through Inuktitut. These key dimensions to self-determination in research are essential to the future, yet they are still not harnessed well within the Eurocentric formal educational system or school science.

12.4 Discussion

In this chapter, I presented two very different projects: first, high school science clubs exploring biodiversity in the city and science in an urban garden and, second, a water stewardship project in Inuit Nunangat. I shared examples of how these places acted pedagogically, offering unique opportunities at the intersection of formal and informal science. These intersections provide insights into the ways in which informal science can support students and youth in building relationships with science, with place, and with each other. Each of the three cases revealed new visions of students as learners who could become stewards of their communities (Cases 1 and 2) or their land, language, culture, and history (Case 3).

In Project 1, the participating students reconnected with nature and thereby developed new or deeper understandings of some science concepts such as biodiversity and endangered species in place, particularly as entangled with their own lives. In the garden, they had embodied experiences with nature guided by an adult who understood where they were coming from and who respected them as individuals. The project underlines the importance of informal science practices that allow for building deep relations among members over time. Furthermore, such relations then became generative of science learning, offering youth opportunities for meaningful forms of engagement with science grounded in authentic experiences and respectful relations with nature in place—in the first case, through explorations of their neighborhood park and nature venues in the city; in the second case, through engagement with nature through gardening and other camp activities.

Participation in these activities deepened the affective connections among the youth resulting in a generative affective atmosphere that took hold over time and emerged from such long-term relations (Ehret et al., 2018). The latter points to the importance of affective placemaking and dignity-affirming learning and becoming in and through science (Nasir et al., 2021). The two cases speak well to the uplifting relational processes among youth, educators, and more-than-human beings that quality programs may support but should be harnessed by researchers as the units of analysis through which to understand the educational potential of informal science programming. Such a lens may help us move toward deeper appreciation of the multispecies community building that can emerge from sameness (e.g., the deep relations among learners and educators), yet that are then also generative of explorations of difference through unknown plants and human-material-nature relations, as the tasting of blackberries or the eco-art project in Case 2 illustrate (Hecht & Nelson, 2021). Unfortunately, much informal science education programming is often short-lived.

In Project 2, participating youth reconnected with the land, their culture, and history through respectful relationships that were brought alive through a practice of science deeply seated in Inuit knowledge systems. As noted by Lipe (2019), "Indigenous knowledge systems are participatory and have been created through lived and practiced applications with nature; they also require knowledge in order to have an application and use" (p. 461). The project offered youth opportunities to engage in environmental stewardship that was meaningful, locally driven, and deeply grounded in traditional Inuit knowledge. The project supported what Alfred (2009) described as resurgence, "the re-creation of sound communities, individual empowerment, and the re-establishment of relationships based on traditional values" (p. 60).

A "form of reattachment of our bodies to our lands" (Simpson, 2017, p. 44) was sought in both projects but was grounded epistemologically somewhat differently. In Cases 1 and 2, the high school students had rich and complex histories: many speaking three languages and navigating among cultural discontinuities between home and school, where they were expected to perform well because their family's immigration implied high hopes for their educational achievement and future. Within that complex fabric, the science clubs became a place where they could engage with science in ways complementary to what they learned in school. The clubs included fieldtrips and activities that built on their interests, contributed to the development of science literacy in meaningful ways, and connected to their lives and their communities. Such learning opportunities can make science meaningful and locally relevant to students who may lack access to science elsewhere or have become disenfranchised from science (Calabrese Barton & Tan, 2010; National Research Council, 2009). Such opportunities are consequential in that they accumulate in "highly personalized ways" of knowing and being in science, over time, and constitute youths' future lives in important ways (Allen & Peterman, 2019, p. 19). Students learned to negotiate meanings and were guided by the instructors to resolve disagreements or make editorial decisions. Most importantly, Cases 1 and 2 embodied an epistemic grounding of students and student learning and becoming that was uplifting and strength-based, supportive of multiple forms of engagement while embracing diversity and multiplicity with a focus on the circulation of bodies and materials in and through nature. The bodies of learners became entangled with nature and its materials in complex ways over time; once I focused on the interconnections that became uplifting to the students (Hecht & Nelson, 2021), their reclaiming of meaningful relationships became evident.

Case 3 was deeply seated in an Indigenous epistemology and offered youth rich opportunities to develop a repertoire of practices that contribute to their well-being both locally and globally as individuals who came to see themselves as proud Inuit invested in lifelong learning, stewards of the land, and Inuit ways of knowing. The building of such new relationships with science, place, and each other through engagement in learning that was meaningful and empowering to the participating youth implied sustained social relationships among youth and mentors (Barron et al.,

2014) and the emergence of deep emotional affinity (Ehret et al., 2018). Youth became involved in something that was meaningful and driven by a clear purpose; they put science to use and did it from the heart. Being on the land for sampling, observing, and learning from Elders was uplifting for the youth; it became a means to make space for Inuit knowledge systems and relationality that is embedded within and part of lived science.

In Case 2, the camp instructors in the Botanical Garden lacked history with the students whom we brought into the garden and could not relate to them in any meaningful way at first. They had no shared history of relationships with our students who were quite different from the campers they were used to, the latter being primarily upper-middle class white children with no history of immigration. Formal education steeped epistemologically in the objectification of individuals and narrowly defined ways of knowing and being in science typically makes the ways of learning previously described in all three cases invisible. A relational epistemology, however, values difference and multiplicity; the subjective and emotional and flourishing together is understood as essential to building possible futures. Lacking such an epistemological lens grounded in interrelationships, the instructors in Case 2 could not possibly appreciate the richness that existed within that space among the students and the campers who all brought with them different "entanglements and configurations of multiple trajectories, multiple histories [or the] thrown togetherness" that made for a garden program that could have been understood as a product of interrelationships and as always under construction as these people, voices, and histories become entangled in new ways through the shared activities (Massey, 2005, p. 148).

In Case 3, a relational yet also holistic epistemology ensured "cultural coherence" in ways water stewardship was practiced and thereby also became culturally sustaining and revitalizing (Tagalik et al., 2023). The project countered environmental dispossession and as such is "less a retreat to nature than a return to familiar ontologies and places" (Robertson & Ljubicic, 2019, p. 549). Accordingly, the project offered opportunities for the "*doing* of Inuit Knowledge (IK)" (p. 553) linked to the land and community well-being. The youth who participated in the project came to live, appreciate, and rebuild relationships that had been broken through colonisation over time. Rebuilding relationships and a relational ontology are key to community well-being and health; it naturally calls for and is deeply grounded in self-determination and political autonomy of Inuit in Inuit Nunangat (Tagalik et al., 2023). It implies a conscious move away from the domination of Western science over Indigenous knowledge or an add-on approach, which is too often still the case within formal education. The program was deeply seated and committed to an Indigenous epistemology. In fact, the water stewardship project can be understood as a political act in that it challenged who was in charge of the research and the epistemology driving it.

As noted by Johnson and Nelson-Barber (2019), such culture-based education programs are crucial for self-determination and contribute in important ways to language revitalization and the sustainability of Indigenous knowledge systems. They are also committed to "quality of learning through an emphasis on personal relationship with the learner and by relying upon a deep trust in the natural unfolding process of human development and learning as part of life itself" (pp. 1348–1349).

12.4.1 *Implications for Social Justice-Driven Science Education*

Social justice issues surface once one begins to study learning and being in science through a lens of relationships that decentres a focus on the objective and individual. As the three cases show, a relational perspective is generative of new stories of learning and becoming. Such stories have much potential in challenging histories of privilege and power in science education, addressing what counts as evidence of learning and becoming at the intersections of formal and informal science. By creating a space for dialogue among epistemologies that uplift silenced or unnoticed science practices, the cases speak to social justice and the urgent need to achieve equitable learning opportunities and outcomes for all students (Adams et al., 2020). By attending to "cultural, epistemic and linguistic pluralism" (Kayumova & Dou, 2022, p. 1), the cases highlight shifts that are needed to focus on "care, interdependence and relational accountability [, understood as] different ethico-political forms of action and thought in science education" (p. 2) and essential to inclusion and social justice driven-science practices. In Case 3 particularly, reflections from the participants speak to the rebuilding of relationships with land, with Elders, and with cultural practices and knowledge systems not taught within the formal educational system.

12.5 Concluding Remarks

Through engagement with data from three very different informal science practices, I have underlined the richness of learning and becoming in science that such practices can support. Informal settings offer challenges and opportunities for innovative approaches and activities that matter to the students and youth involved. As such, informal science settings can be important spaces of learning and becoming in science that complement formal settings. The cases were situated in two very different regions of Canada, showing how young people can engage with science education grounded in different epistemologies. By applying a relational lens to learning and becoming in three different cases, I challenge narrowly defined forms of science education steeped in Western epistemology, which does not speak to diverse learners across spaces of learning and becoming. I suggest that relational epistemologies are crucial to viewing learning and becoming as entangled instead of static and as taking form at the intersections of informal and formal education. Building new relationships through informal science as illustrated in the two projects is an invitation to all science educators to more deeply and critically engage with the taken-for-granted epistemologies (ways of knowing), ontologies (concepts), and axiologies (values and ends that drive the former) constitutive of learning and becoming in science for all. Coming to know science is about embracing multiplicity and complexity while engaging humbly with justice-driven and grounded agendas—essentially a call for a new ecology of thought and purpose for science education.

Acknowledgements This chapter was written in memory of Eric, a youth participant in Case 3, who tragically lost his life while working on a fishing vessel in the summer of 2021 in the area of Mittimatalik (Pond Inlet). He had been working with the fishing company as a crew member for 3 years and was much appreciated. His contributions to the water stewardship project extended beyond his work in Pond Inlet as documented here, implying some science training in one of the partner universities and contribution to an oral presentation at Arctic Net's annual Scientific Meeting. To highlight his engagement and gifts, his voice is shared in respectful ways and with much humility.

I thank all the youth and adults who made both projects possible. In Cases 1 and 2, the "we" includes two research assistants, Gwénaëlle Journet and Issac Hebert, in each club, who organized and ran the clubs, with a third research assistant, Émilie Boulanger, collecting the data. The project was funded through a 3-year grant from FRQ-SC. https://frq.gouv.qc.ca/app/uploads/2021/07/prs_rahmj_annexe_club-scientifique.pdf.

For Case 3, I thank the project leaders, Tim Soucie Anaviapik and Trevor Arreak, all the youth and families in Mittimatalik (Pond Inlet) as well as ARCTIConnexion for their support throughout this project. The project was funded through a 3-year grant from FRQ-SC. https://frq.gouv.qc.ca/app/uploads/2021/04/rahmj_prs_projet-collaboratif-avec-les-jeunes-inuits_rapport.pdf.

References

Adams, J. D., Das, A., & Kim, E.-J. A. (2020). The crit-trans heuristic for criticalizing STEM education: Youth and educators as participants in the world. In S. R. Steinberg & B. Down (Eds.), *Sage handbook of critical pedagogies* (Vol. 3, pp. 1497–1507). Sage. https://doi.org/10.4135/9781526486455

Alfred, T. (2009). *Peace, power, righteousness: An indigenous manifesto* (2nd ed.). Oxford University Press.

Allen, S., & Peterman, K. (2019). Evaluating informal STEM education: Issues and challenges in context. *New Directions for Evaluation, 2019*(161), 17–33. https://doi.org/10.1002/ev.20354

Anaviapik-Soucie, T., Arreak, T., L'Hérault, V., & Elders of Pont Inlet. (2015). *Building capacity to monitor fresh water quality in Pond Inlet. Historical and cultural perspective on water uses—Risks to human health-climate and environmental changes = recommendations for adaptation* [Unpublished manuscript].

Bang, M. E., Marin, A., & Medin, D. (2018). If Indigenous peoples stand with the sciences, will scientists stand with us? *Daedalus, 147*(2), 148–159. https://doi.org/10.1162/DAED_a_00498

Barron, B., Gomez, K., Pinkard, N., & Martin, C. K. (Eds.). (2014). *The digital youth network: Cultivating digital media citizenship in urban communities*. MIT Press. https://doi.org/10.7551/mitpress/9240.001.0001

Cajete, G. A. (2000). *Native science: Natural laws of interdependence*. Clear Light Publishers.

Calabrese Barton, A., & Tan, E. (2010). We be burnin'! Agency, identity, and science learning. *Journal of the Learning Sciences, 19*(2), 187–229. https://doi.org/10.1080/10508400903530044

Ehret, C., Boegel, J., & Manuel-Nekouei, R. (2018). The role of affect in adolescents' online literacies: Participatory pressures in BookTube culture. *Journal of Adolescent & Adult Literacy, 62*(2), 151–161. https://doi.org/10.1002/jaal.881

Hecht, M., & Nelson, T. (2021). Youth, place, and educator practices: Designing program elements to support relational processes and naturalist identity development. *Environmental Education Research, 27*(9), 1401–1420. https://doi.org/10.1080/13504622.2021.1928608

Holland, D. C., Lachicotte, W. S., Jr., Skinner, D., & Cain, C. (1998). *Identity and agency in cultural worlds.* Harvard University Press.

Johnson, Z., & Nelson-Barber, S. (2019). Always alert, always agile: The importance of locally researching innovations and interventions in Indigenous learning communities. In E. A. McKinley & L. T. Smith (Eds.), *Handbook of Indigenous education* (pp. 1347–1384). Springer. https://doi.org/10.1007/978-981-10-3899-0_73

Jordan, B., & Henderson, A. (1995). Interaction analysis: Foundations and practice. *Journal of the Learning Sciences, 4*(1), 39–103. https://doi.org/10.1207/s15327809jls0401_2

Kayumova, S., & Dou, R. (2022). Equity and justice in science education: Toward a pluriverse of identities and onto-epistemologies. *Science Education, 106*(5), 1097–1117. https://doi.org/10.1002/sce.21750

Lave, J., & Wenger, E. (1991). *Situated learning: Legitimate peripheral participation.* Cambridge University Press.

Lipe, D. (2019). Indigenous knowledge systems as the missing link in scientific worldviews: A discussion on Western science as a contested space. In H. Tomlins-Jahnke, S. D. Styers, S. Lilley, & D. Zinga (Eds.), *Indigenous education: New directions in theory and practice* (pp. 453–481). University of Alberta Press.

Marin, A. M. (2020). Ambulatory sequences: Ecologies of learning by attending and observing on the move. *Cognition & Instruction, 38*(3), 281–317. https://doi.org/10.1080/07370008.2020.1767104

Massey, D. (2005). *For space.* Sage.

Medeiros, A. S., Wood, P., Wesche, S. D., Bakaic, M., & Peters, J. F. (2017). Water security for northern peoples: Review of threats to Arctic freshwater systems in Nunavut, Canada. *Regional Environmental Change, 17*(3), 635–647. https://doi.org/10.1007/s10113-016-1084-2

Nasir, N. S., Lee, C. D., Pea, R., McKinney de Royston, M. (2021). Rethinking learning: What the interdisciplinary science tells us. *Educational Researcher, 50*(8), 557–565. https://journals.sagepub.com/doi/abs/10.3102/0013189X211047251

National Research Council. (2009). *Learning science in informal environments: People, places, and pursuits* [Consensus Study Report]. National Academies Press. https://doi.org/10.17226/12190

Nxumalo, F. (2020). Situating Indigenous and Black childhoods in the Anthropocene. In A. Cutter-Mackenzie-Knowles, K. Malone, & E. Barratt Hacking (Eds.), *Research handbook on childhoodnature: Assemblages of childhood and nature research* (pp. 535–556). Springer. https://doi.org/10.1007/978-3-319-67286-1_37

Rahm, J., Fahrni, L., & Touioui, F. (2023). A mobile theory of learning and identity in and through relations of dignity: A research framing for research outside the classroom. In P. G. Patrick (Ed.), *How people learn in informal science environments* (pp. 163–184). Springer. https://doi.org/10.1007/978-3-031-13291-9_9

Rahm, J., Gonsalves, A. J., & Lachaîne, A. (2022). Young women of color figuring science and identity within and beyond an afterschool science program. *Journal of the Learning Sciences, 31*(2), 199–236. https://doi.org/10.1080/10508406.2021.1977646

Rieber, R. W., & Carton, A. S. (Eds.). (1987). *The collected works of L.S. Vygotsky: Problems of general psychology, including the volume thinking and speech* (Vol. 1). Plenum Press/Springer. https://doi.org/10.1007/978-1-4613-1655-8.

Robertson, S., & Ljubicic, G. (2019). *Nunamii'luni quvianaqtuq* (It is a happy moment to be on the land): Feelings, freedom and the spatial political ontology of well-being in Gjoa Haven and Tikiranajuk, Nunavut. *Environment & Planning D: Society & Space, 37*(3), 542–560. https://journals.sagepub.com/doi/abs/10.1177/0263775818821129

Simpson, L. B. (2017). *As we have always done: Indigenous freedom through radical resistance.* University of Minnesota Press.

Sobel, D. (2012, July-August). Look, don't touch. *Orion.* https://orionmagazine.org/article/look-dont-touch1/

Tagalik, S., Baker, K., Karetak, J., & Rahm, J. (2023). Rebuilding relations and countering erasure through community-driven and owned science: A key tool to Inuit self-determination and social transformations. *Journal of Research in Science Teaching, 60*(8), 1697–1722. https://doi.org/10.1002/tea.21881

Takeuchi, M. A., Vadeboncoeur, J. A., Krishnamoorthy, R., Hladik, S., Rahm, J., Kim, B., & Clark, D. B. (2023). Refiguring and transforming the learning sciences: A dialogical field-in-the-making. In M.-C. Shanahan, B. Kim, M. A. Takeuchi, K. H. Koh, A. P. Preciado-Babb, & P. Sengupta (Eds.), *The learning sciences in conversation: Theories, methodologies, and boundary spaces* (pp. 255–266). Routledge.

Tan, E., & Calabrese Barton, A. (2023). *Teaching toward rightful presence in middle school STEM.* Harvard Education Press.

Taylor, C. A. (2016). Edu-crafting a cacophonous ecology: Posthumanist research practices for education. In C. A. Taylor & C. Hughes (Eds.), *Posthumanist research practices in education* (pp. 5–24). Palgrave Macmillan. https://doi.org/10.1057/9781137453082_2

Tuck, E., & McKenzie, M. (2015). *Place in research: Theory, methodology, methods.* Routledge.

Wilson, S., & Hughes, M. (2019). Why research is reconciliation. In S. Wilson, A. V. Breen, & L. DuPré (Eds.), *Research and reconciliation: Unsettling ways of knowing through Indigenous relationships* (pp. 5–19). Canadian Scholars.

Jrène Rahm, BA (McGill University), MA (University of Colorado at Boulder, USA), PhD (University of Colorado, USA), is an Educational Psychologist and Full Professor in the Faculty of Education and Department of Educational Psychology at the Université de Montréal, Canada. Prior to this, she worked as a Professor at the University of Northern Colorado in the USA. Her work entails projects conducted together with ethnically diverse and Indigenous youth and their communities in diverse educational settings exploring learning and identity in informal science. Through multi-sited ethnographies and longitudinal studies grounded in sociocultural and critical theory, her recent work focuses on the entanglements of bodies with materials, affect and dignity affirming and uplifting educational opportunities in science deeply committed to and driven by equity and social justice, generative of new possibilities and futures.

Chapter 13
Tracking Topography in Canadian Science Education Research

Dawn Wiseman

13.1 Introduction

Canadian science teacher educators and science education researchers are often asked by colleagues from other nation-states to speak to/about science education in Canada as a whole. Perhaps it is a polite question to initiate conversation, perhaps it is interest brought about by Canada's competitiveness in international student assessments in science (Council of Ministers of Education, Canada, 2023), or perhaps it is an attempt to understand Canadian science education and science education research in comparison and/or relation to where the questioner is from. No matter what the impetus, individually and collectively (e.g., Bowen et al., 2023; Hart, 2002) my colleagues and I respond as best we can, typically beginning by pointing out that—except in some very specific cases (see Phillips, 2019; Wiseman, 2016) in Canada—education is a provincial/territorial responsibility with no federal oversight (Government of Canada, 2024). That is, while Canadian science teacher educators and science education researchers meet and share thinking on a national level, how (science) education is framed in Nunavut, or Prince Edward Island, or Saskatchewan is specific to that province or territory. In Canadian science education, jurisdictional differences manifest in several ways. Sometimes they appear as relatively straightforward reflections of context; for example, in different provinces/territories, teachers exploring energy may choose to do so via local production processes such as fossil fuels in Alberta or tidal hydro-electric in Nova Scotia. Sometimes the differences are more deeply rooted in cultural contexts or in the foregrounding of First People's ways of knowing, being, and doing; for example, Nunavut's commitment to the principles

With apologies to Cynthis Chambers.

D. Wiseman (✉)
Bishop's University, Sherbrooke, QC, Canada
e-mail: dwiseman@ubishops.ca

C. D. Tippett and T. M. Milford (eds.), *Seeing Science Through the Eyes of Canadian Teachers and Learners*, Contemporary Trends and Issues in Science Education 54,
https://doi.org/10.1007/978-3-031-88785-7_13

of Inuit Qaujimajatuqangit (Wiseman & Kreuger, 2019) or the Northwest Territories creation of experiential senior high school science courses grounded in terrestrial, marine, and fresh water systems (Northwest Territories Education, Culture, & Employment, 2006) rather than the usual disciplinary areas of biology, chemistry, and physics. In all cases, these differences impact the kind of science education and science education research that occurs.

In an attempt to wrestle with the frequent inquiry into how we explain ourselves as science educators and science education researchers in Canada, the editors of this collection note its intent is to bring together contributions that demonstrate the breadth of science education research being done across the country. While acknowledging the heterogeneity at play (Chap. 1), they suggest there is something distinctly Canadian about science education research done within the geographic boundaries that define Canada. They point to topics taken up by researchers at Canadian institutions that, while not unique within the field of science education research internationally, seem more prevalent in multiple places of Canada. That difference in pattern from what might be considered the global norm indicates something interesting might be occurring—something demonstrable, or at least trackable, but also something challenging to pin down and define.

Considering what might be going on in Canadian science education research and how to track it, I find myself drawn to Chambers' (1999) vision of a topography of Canadian curriculum theory and theorists. She suggested that by mapping "who we are … [in relation to the physical, sociopolitical, historical, and institutional specificities of the very local places] where we live and work" (p. 144) we might find both what we have in common and "a place to begin the difficult work of reaching into and across the territories of difference" (p. 148). As she underlined, development of such a topography is a "humble … task" (p. 147) that presents several challenges to curriculum theorists, including:

> 1. Creat[ing] curriculum languages and genres that name the sociopolitical, geophysical, and imaginative landscape in which Canadians live now, as well as the landscapes of the past and the future;
>
> 2. Turn[ing] to Canadian scholars, indigenous languages, and traditions for that language and those genres;
>
> 3. Seek[ing] new interpretive tools for understanding what it means to be Canadian and what Canadians might become in the twenty-first century;
>
> 4. Creat[ing] curriculum theory that is written at home but works on behalf of everyone. (p. 137)

Given the quarter century that has elapsed since her initial writing, I suggest another challenge is finding ways to account for how curriculum theory, like topography, shifts and changes in relation to the impact of time. However, the focus of this chapter is not to expand from Chambers' thinking but rather to consider how the chapters in this collection might be viewed together.

I acknowledge that researching and theorizing are not the same and that science education researchers and curriculum theorists may have significantly different underlying philosophies (and sometimes paradigms) to their work. At the same time, I suggest that, in Canada, science education research and researchers are a subset of, or at least analogous to, curriculum theory and theorists because they are responding to the same sort of contextual circumstances. Working from this assumption, I further suggest that the challenges Chambers (1999) outlined with respect to Canadian curriculum theory also apply to Canadian science education research: That who we are and where we find ourselves shapes the science education research that occurs within the physical boundaries of Canada. In what follows, I have used these challenges as a framing device for considering what can be said about the topography of Canadian science education research as presented in this collection. That is, the challenges provide some means of recognizing the specificities of Canadian science education research.

Drawing from a slightly more recent piece by Chambers (2003), I acknowledge there are limits to this approach. I am perhaps structuring the 1999 challenges too much—my first degree is in engineering; I do like a structure. I intend for the challenges to serve more as invitations to attend to while theorizing from, about, and with/in Canada; however, I recognize that I am still not accounting for the full complexity and abundance of Canadian science education research that my peers offer. In addition, I acknowledge that I write from a particular position: I am a white woman from Québec, with connections and relations to my science education peers from sea to sea to sea, but work at a small, regional university in (what is locally) a minority language. I come to science education research from many years of working with and alongside Indigenous educators and schools (e.g., Robinson et al., 2023). I study Canadian science education research with colleagues (e.g., Pegg et al., 2015). From this work, I see research as a deeply relational rather than objective undertaking; an assumption that no doubt impacts my choice of framing here as well as how I read the concepts of Canada and Canadian in this collection. Like Chambers (2003), I often find that available language and standard practice do not allow for expression of the fulsome joy of the research my colleagues and I do, or even the acknowledgement that part of what we do is because science, science education, and science education research is serious fun. So, I invite you to read what follows as a guide in support of further journeying and finding your own pathway through Canadian science education research rather than as a definitive comment on the state of science education research in Canada.

13.2 The First Three Challenges

Chambers' (1999) challenges allow for consideration both of where there is something tangible occurring that suggests (a) who we are and where is here and (b) where there are absences/silences in this regard. She moved beyond an objective observation and description of research/theorizing to explore processes and ethics that shape how research/theorizing is taken up. In considering the science education research within this collection, I sought out conversations between the challenges and the chapters then thought about what they might imply for a topography of Canadian science education research. In this section, I consider the first three challenges. While they are treated separately, there are interconnections and intertwinings among them. I address the fourth challenge in a separate section as a means of moving toward some closing thoughts.

13.2.1 Creating and Writing Science Education Research from This Place

Chambers (1999) described the first challenge as a "nam[ing] of where we are, and what it looks and feels like to be in this place, even when we feel 'out of place'" (p. 144). It is a challenge to not solely locate where research occurs—although such location is important—but to recognize relationality with place as critical to thinking, learning, and research. Within the chapters in this collection, authors take up writing from this place in several ways. In considering the first challenge, I acknowledge that curriculum theorizing and science education research often take on different forms of writing. So, considering Chambers' invocation within the first challenge to engage with, for example, the poetic as a means of experiencing, expressing, and coming to understand, within these chapters the grittiness of a Prairie landscape as a form of curriculum may not necessarily be evident to all readers. At the same time, even though chapters seem grounded in what appears to be more traditional academic approaches to writing, there is no question that their writing engages with the sociopolitical, geophysical, imaginative, historical, and speculative landscapes of Canada.

While the poetic might be challenging to discern, as a displaced Montrealer, I can attest to how Rahm's (Chap. 12) description of the city's botanical gardens and biodiversity centre brought me directly to those places and the learning she describes occurring in them. It is how I know these places to feel and be, and how I can imagine their potential as sites from which science teaching and learning might emerge. That resonance with a place I know in my bones extends to her description of Pond Inlet as an Inuit community struggling with a problem all too common to the Indigenous communities that Canada is mapped onto: Access to safe water supply due to ongoing sociopolitical impacts of colonial policy. Thus, her writing of this place within her contribution provides not only a means for opening issues of equity and social justice, but also provides a topography for reading the research much like we read the land.

Similarly, Wells and Ricketts (Chap. 5) write of the ocean as a particular place and relation within the context of Newfoundland and Labrador. Within their research, standard explorations of the ocean (e.g., albedo, acidification, etc.) occur in a place where most students live close to the Atlantic coast and thus understand the history and impact of biotic and abiotic oceanic changes on the people of the province and their families. While the authors mention northern cod only once in the chapter, given the long history of cod fishing in the province and the impact of its moratorium, there are few Canadian science education researchers who could read the chapter and not understand the implication that ocean literacy is key to the ongoing recovery of the species and the cultural and economic well-being of the people of Newfoundland and Labrador. Political decisions and shared rememberings of cod haunt this chapter and enliven the reasons for such research. Kelly and Pelech (Chap. 11) also write of Newfoundland and Labrador with a more fulsome history of Marystown and the relationships among its people, place, and the invasive European green crab. In their work, topographies of science (what constitutes an invasive species) and science education (provincial curricula) as defined by governments play a key role. And so, the history of the land, air, waters, and relations between human and other-than-human relatives are woven into these research chapters in ways specific to Canada and to Canadian science education and science education research.

There is no question that the constitutional reality of Canada and its division of responsibilities between federal and provincial/territorial governments are a key topographical feature in these chapters. Many authors write their science education research and their place with a foregrounding and/or explicit acknowledgement of provincially produced directives on Kindergarten through Grade 12 (K–12) science curricula (Chaps. 2, 5, 7, & 11), evaluation processes (Chap. 6), and teacher education (Chap. 10). Informed readers can also see how the science education landscape is impacted by curricular policies less explicitly described in the chapters. For example, both Meunier and Bélanger (Chap. 10) and Rahm (Chap. 12) write Québec "from this place, of place, and for this place" (Chambers, 1999, p. 145) through the informal cultural sites that the provincial science curricula encourages teachers to use as locations and resources for teaching and learning (Ministère de l'éducation, 2004, p. 230). The manner in which science education and the science education research landscape is impacted by the complexity and tensions of the federal/territorial split is evident in Raisinganhi's (Chap. 4) exploration of English Language Learners, where she cites the *Common Framework of Science Learning Outcomes, K to 12* (Council of Ministers of Education, Canada, 1997). This nationally negotiated document both (a) laid out basic assumptions about the impetus for and focus of science education in Canada that continue to impact provincial science programs and (b) facilitated movement of young people between different provincial/territorial jurisdictions by defining benchmarks for particular science learning outcomes.

As a whole, the chapters in this collection demonstrate how policy and policy change over time interweave and move to shape science education and science education research at multiple levels—much like water or wind over rock. Additionally, these chapters name the topological location(s) of universities within science education research in Canada through relations to postsecondary learning of science (Chaps. 8 & 9) and places where science teachers are educated (Chap. 3). What emerges in these chapters are examples of institutions of higher learning as places with their own topographies. For Noble (Chap. 9), this place is Royal Roads University and the institutional specificities of its commitment to students learning at a distance with an emphasis on project-based learning. Vergis (Chap. 8) takes a similar approach at the University of Alberta in relation to concepts of evidence in undergraduate chemistry. Importantly, Macleod and Kraglund-Gauthier (Chap. 3) write about changes to this place in the face of the global through the impact of the pandemic on science teacher education in a very local program, demonstrating the importance of writing from this place, "even we [and everyone else] feel 'out of place'" (Chambers, 1999, p. 144). The earth-shattering realities of a global impact play out even in small, local spaces. Canadian science education researchers write from place in ways that allow local, provincial, and national topographies of the field to emerge—sometimes explicitly, sometimes less so.

13.2.2 Turning to a Science Education Research Language of Our Own

As Chambers (1999) laid out, "the second challenge will be to find and create a language of our own, and to turn to our own for that language" (p. 145) through conversations with each other and the place/s where we live, play, and do research. The contributions to this collection are certainly in conversation with place as noted in the previous section. All contributions in the collection (some more than others) cite other work/research/policy by people located in Canadian K–12, postsecondary, nongovernmental, and governmental institutions. So, the language of curriculum documents and concerns at local, provincial/territorial, and national levels play out throughout the collection. For example, Raisinganhi (Chap. 4) engages deeply with commitments in the British Columbia K–12 program to social justice, diversity, and Indigenous worldviews alongside work by other Canadian researchers on the role, place, and impact of multiculturalism, difference, diversity, and immigration in science teaching and learning. Kelly and Pelech (Chap. 11), similarly, build from other Canadian researchers' conceptions of ecological education, place-based learning, and learning from and with land.

Rahm (Chap. 12) perhaps goes deeper than most others in the collection in taking up the difficult work of language that Chambers (1999) noted "begin[s] by learning to hear each other" (p. 146). Along with a submission written in English (not her first language), Rahm takes up Chambers' push for scholars to "move beyond" (p. 146) official languages, curricula, etc. to embrace the diversity of languages that circulate in Canada: Indigenous languages rooted in the land for millennia and more recently arrived languages of the global diaspora found in the everyday places where the fullness of life, living, and relationships with the world are at play. Her contribution occurs in conversation with work by Indigenous scholars to examine learning in contexts beyond the classroom and official curricula.

Chambers (1999) underlined that a potential outcome of developing a language of our own is "a hybrid trade language of curricular and linguistic multiplicity [that] could offer Canadian curriculum scholars the possibility of both creating and locating a curricular landscape of our own" (p. 146). Within the confines of the science education research presented in this collection, this kind of emerging hybrid language is evident in only a few cases; however, in taking a broader view that includes interpretive tools, it perhaps becomes more evident.

13.2.3 Seeking Interpretative Tools of Our Own

Chambers' (1999) third challenge is "to seek out or create interpretive tools that allow Canadian curriculum theorists to write and interpret who Canadians are, what we know, and where we want to go, all the while remaining cognizant of an important truism: There will be no single answer to these questions" (p. 146). As with language, she suggested these tools be sought "within the physical and imaginary landscape of Canada" (p. 147) and that they might be hybrid; that is, they may draw on traditions and heritages from other places but "be made anew, infinitely re-created in and with a land and people that seem to refuse easy definition" (Chambers, 2003, p. 246). Through a reading and re-reading of the chapters in this collection, the Canadian-ness of the whole and parts become most evident in how interpretative tools theorized by Canadian curriculum scholars (e.g., Aoki, 2004; Chambers, 1999, 2003; Connelly & Clandinin, 1990; Jardine, 2006; van Manen, 1982, 1991) underlie texts focused on science education research, often implicitly. I point to only some examples in what follows.

Perhaps the most evident is Aoki's (2004) idea of curriculum-as-plan and curriculum-as-lived-experience, where the tensions between official conceptions of curricula defined by governments and institutions and what happens in the messy experience of living together in teaching and learning can either shut down or open up educational possibilities. These tensions and the ability to dwell between them so that learning emerges are present in authors' considerations of how curriculum, teaching, and learning can come together in relation to land (Chaps. 5, 11, & 12) and challenging practices such as assessment of laboratory reports (Chap. 6), concepts of evidence (Chap. 8), integration of local/cultural resources (Chaps. 7 & 10), online inquiry (Chaps. 3 & 9), navigating plurilingualism (Chap. 4), and implementing curriculum change (Chap. 2).

Canadian interpretative frameworks, such as Jardine's (2006) uptake of Gadamerian hermeneutics, can also be read into this collection. Several chapters take up the "fecundity of the individual case" (p. 151) or how it is possible to make meaning from specific, local instances. For example, to support science educators in their use of media sources in the classroom, Reis et al. (Chap. 7) present the case of a science teacher using scientific news in the classroom. While the news story, classroom setup, and location are unique, their description of the teacher's practice of drawing student attention to aspects of the news article, helping them to connect information with prior learning, and bridging every day and scientific language will support science educators in recognizing and building on their own practices.

Several chapters focus on the sharing of stories, vignettes, and experience that seem to draw on interpretative methods, if not methodologies, as taken up in Connelly and Clandinin's (1990) narrative inquiry and/or van Manen's (1982, 1991) phenomenology. Notable is MacLeod and Kraglund-Gauthier's (Chap. 3) narrative conversation that explores the experience of teaching science under the emergency remote conditions of the pandemic. Other chapters (5 & 11) cross interpretative lines engaging with ecological thinking per Greenwood (2013) and deeper conceptions of relationship and kinship put forward by Indigenous and Métis scholars such as Donald (2013) and Todd (2014, 2022), respectively. Todd is not technically a curriculum scholar, but her work raises questions regarding where understanding is located and whose knowledge is privileged, which are very much questions of curriculum. These chapters extend thinking about where and how science teaching and learning might be taken up as well as what relationships are attended to and privileged within science teaching and learning.

The interpretive tools pointed to above are generally not explicit in the chapters in this collection. In fact, each chapter outlines methods and methodologies of its own: case study, vignette methodology, discourse analysis, etc. But, in my reading of these chapters, it is as if the interpretative approaches and commitments of curriculum studies as taken up by Canadian curriculum scholars are implicit to the nature of science education research in the same context. These interpretative approaches seem to circulate in the background of Canadian science education research and influence "who we are" (Chambers, 1999, p. 144) in relation to the physical, sociopolitical, historical, and institutional specificities of the very local places "where we live and work" (p. 144), a part of the ethos in which we find ourselves.

13.3 Creating a Topography of Canadian Science Education Research

One hallmark of a strong collection of work on a topic is how it invites multiple readings, both in terms of returning to the work and in how it might be interpreted. Multiple readings reflect topography as process, in particular, how a field is characterized by its key features and shaped by its interactions over time with its inhabitants and influences from beyond its own bounds. This collection provides a clear reflection of the topography of Canadian science education research despite the relatively small number of studies presented. Key features include engagement with place and the specificity of the local within broader conceptions of Canada. They reflect the tension of provincial/territorial responsibilities. They highlight relationships through work that is overwhelmingly in the service of young people/learners and their teachers. In this commitment to relationship, these features demonstrate how Canadian science education research is caught up in questions of curriculum (mostly as-lived-experience; Aoki, 2004) and praxis. It is here where I wonder if the overwhelming choice of qualitative methods employed in the studies in this collection is an indication of how Canadian science education research emerges from the tensions of more general, global research practice into science education and the strong legacy of curriculum studies in Canada. Perhaps this methodological choice is how we might both define "who we are [and] where we live and work" (Chambers, 1999, p. 144). I look forward to other people's thoughts on the matter.

In closing, like Chambers (2003), I must acknowledge "a profound sense of obligation" (p. 221) to the readers of this chapter and to the authors and editors of this collection. Canada is a large place made of small communities; the science education research community is no exception. It is a true honour to be a member of this community and to be asked to engage with the chapters in this collection.

References

Aoki, T. T. (2004). Teaching as indwelling between two curriculum worlds (1986/1991). In W. F. Pinar & R. L. Irwin (Eds.), *Curriculum in a new key: The collected works of Ted T. Aoki* (pp. 159–165). Lawrence Erlbaum Associates. https://doi.org/10.4324/9781410611390

Bowen, G. M., Wiseman, D., Shanahan, M. C., Khan, S., Gonsalves, A., Sengupta, P., Simms, W., Knoll, E., & Carter, A. (2023). STEM in Canadian teacher education: An overview. In S. M. Al-Balushi, L. Martin-Hansen, & Y. Song (Eds.), *Reforming science teacher education programs in the STEM era: International and comparative perspectives* (pp. 53–70). Springer. https://doi.org/10.1007/978-3-031-27334-6_4

Chambers, C. (1999). A topography for Canadian curriculum theory. *Canadian Journal of Education/Revue Canadienne De l'éducation, 24*(2), 137–150. https://journals.sfu.ca/cje/index.php/cje-rce/article/view/2749

Chambers, C. (2003). "As Canadian as possible under the circumstances": A view of contemporary curriculum discourses in Canada. In W. F. Pinar (Ed.), *International handbook of curriculum research* (pp. 221–252). Lawrence Erlbaum Associates. https://doi.org/10.4324/9781410607645

Connelly, F. M., & Clandinin, D. J. (1990). Stories of experience and narrative inquiry. *Educational Researcher, 19*(5), 2–14. https://doi.org/10.3102/0013189x019005002

Council of Ministers of Education, Canada. (1997). *Common framework of science learning outcomes, K to 12: Pan-Canadian protocol for collaboration on school curriculum for use by curriculum developers*. https://science.cmec.ca/framework/

Council of Ministers of Education, Canada. (2023). *Measuring up: Canadian results of the OECD PISA 2022 study*. https://www.cmec.ca/Publications/Lists/Publications/Attachments/438/PISA-2022_Canadian_Report_EN.pdf

Donald, D. (2013). Foreword. In A. Kulnieks, D. Roronhiakewen Longboat, & K. Young (Eds.), *Contemporary studies in environmental and Indigenous pedagogies: A curricula of stories and place* (pp. vii–viii). Sense. https://doi.org/10.1007/978-94-6209-293-8

Government of Canada. (2024, December 27). *Learn about education in Canada*. https://www.canada.ca/en/immigration-refugees-citizenship/services/new-immigrants/new-life-canada/education.html

Greenwood, D. A. (2013). A critical theory of place-conscious education. In R. B. Stevenson, M. Brody, J. Dillon, & A. E. J. Wals (Eds.), *International handbook of research on environmental education* (pp. 93–100). Routledge. https://doi.org/10.4324/9780203813331

Hart, P. (2002). Environment in the science curriculum: The politics of change in the Pan-Canadian science curriculum development process. *International Journal of Science Education, 24*(11), 1239–1254. https://doi.org/10.1080/09500690210137728

Jardine, D. W. (2006). "The fecundity of the individual case": Considerations of the pedagogic heart of interpretive work. In D. W. Jardine, S. Friesen & P. Clifford (Eds.), *Curriculum in abundance* (pp. 151–168). Lawrence Erlbaum Associates. https://doi.org/10.4324/9780203826133

Ministère de l'éducation. (2004). *Québec education program, secondary school education, cycle 1*. https://cdn-contenu.quebec.ca/cdn-contenu/education/pfeq/secondaire/PFEQ-secondaire-premier-cycle-AN.pdf

Northwest Territories Education, Culture, & Employment. (2006). *Experiential science 10–20–30*. https://www.ece.gov.nt.ca/sites/ece/files/resources/experimental_science_10_-_30.pdf

Pegg, J., Wiseman, D., & Brown, C. (2015). Conversations about science education: A retrospective of science education research in CJSMTE. *Canadian Journal of Science, Mathematics, & Technology Education, 15*(4), 364–386. https://doi.org/10.1080/14926156.2015.1093202

Phillips, R. S. (2019). A commentary: Education in Canada—Does anyone read our constitution? *Brock Education Journal, 28*(2), 4–16. https://doi.org/10.26522/brocked.v28i2.590

Robinson, L., West, K., Daoust, M., Sylliboy, S., Lafferty, A., Wiseman, D., Lunney Borden, L., Ghostkeeper, E., Glanfield, F., Ribbonleg, M., & Bernard, K. (2023). When mathematics has spirit: Aki Chike Win. *ZDM-Mathematics Education, 55*(6), 1053–1065. https://doi.org/10.1007/s11858-023-01482-7

Todd, Z. (2014). Fish pluralities: Human-animal relations and sites of engagement in Paulatuuq, Arctic Canada. *Études/inuit/studies, 38*(1–2), 217–238. https://doi.org/10.7202/1028861ar

Todd, Z. (2022). Fossil fuels and fossil kin: An environmental kin study of weaponized fossil kin and Alberta's so-called "energy resources heritage." *Antipode*. Advance online publication. https://doi.org/10.1111/anti.12897

van Manen, M. (1982). Phenomenological pedagogy. *Curriculum Inquiry, 12*(3), 283–299. https://doi.org/10.1080/03626784.1982.11075844

van Manen, M. (1991). *The tact of teaching: The meaning of pedagogical thoughtfulness*. Althouse Press.

Wiseman, D. (2016). *Acts of living with: Being, doing, and coming to understand Indigenous perspectives alongside science curricula* [Unpublished doctoral dissertation, University of Alberta]. https://era.library.ualberta.ca/files/ctt44pm87m/Wiseman_Dawn_201603_PhD.pdf

Wiseman, D., & Kreuger, J. (2019). Science education in Nunavut: Being led by Inuit Qaujimajatuqangit. In C. D. Tippett & T. M. Milford (Eds.), *Science education in Canada: Consistencies, commonalities, and distinctions* (pp. 287–310). Springer. https://doi.org/10.1007/978-3-030-06191-3_14

Dawn Wiseman (she/her), BEng (Concordia University), MA (Concordia University), PhD (University of Alberta), is an Associate Professor and Chair in the School of Education at Bishop's University (Sherbrooke, QC/Abenaki territory). She comes to academia after more than 25 years of work in STEM education alongside Indigenous communities, first through the Native Access to Engineering Program at Concordia University (Montreal, QC/Haudensosaunee and Anishnabe territory) and then with Mount Pleasant Educational Services (Kanien'kéha territory of Kahnawà:ke). Her teaching and research focus on the manner in which Indigenous and Western ways of knowing, being, and doing circulate together in STEM education, student-directed STEM education, and teaching and learning in times of intersecting global crises.

Index

C. D. Tippett and T. M. Milford (eds.), *Seeing Science Through the Eyes of Canadian Teachers and Learners*, Contemporary Trends and Issues in Science Education 54,
https://doi.org/10.1007/978-3-031-88785-7

The manufacturer's authorised representative in the EU is Springer Nature Customer Service Centre GmbH, Europaplatz 3, 69115 Heidelberg, Germany. If you have any concerns regarding our products, please contact ProductSafety@springernature.com

Printed and bound by CPI Group (UK) Ltd, Croydon, CR0 4YY

01/10/2026

02225961-0009